The Uses and Abuses of Humour in Social Work

In recent years, social work academics and practitioners have highlighted the need to 're-claim' methodologies which unpack creativity and resourcefulness. In the bleakest of times many turn to humour to survive. Making a unique contribution to social work thought, this is the first book to focus exclusively on humour use in relation to social work.

Over eight chapters Jordan covers a range of examples of social work humour, using examples from practice, fiction and research. He concludes that social work has a complex relationship with humour and that humour has an important role in social work as it enables social workers to hold contradictory views. It also allows society to manage its ambivalent and contradictory view of social work.

Aimed at academics, students and social work professionals, this book explores social work's sometimes uneasy relationship with humour. It will be of interest to anyone with an academic interest in humour.

Stephen Jordan worked as a social worker for 19 years and now works for the University of Essex where he lectures in social work.

T0174902

Routledge Advances in Social Work

Consciousness-Raising
Critical Pedagogy and Practice for Social Change
Nilan Yu

Participatory Pedagogic Impact Research
Co-production with Community Partners in Action
Mike Seal

Forthcoming

Intersectionality in Social Work
Activism and Practice in Context
Edited by Suryia Nayak and Rachel Robbins

Conversation Analysis for Social Work
Talking with Youth in Care
Gerald de Montigny

Older Lesbian, Gay, Bisexual and Trans People
Minding the Knowledge Gap
Andrew King, Kathryn Almack, Yiu-Tung Suen and Sue Westwood

Visual Communication for Social Work Practice
Power, Culture, Analysis
Sonia Magdalena Tascon

Art in Social Work Practice
Theory and Practice: International Perspectives
Ephrat Huss and Eltje Bos

The Uses and Abuses of Humour in Social Work
Stephen Jordan

www.routledge.com/Routledge-Advances-in-Social-Work/book-series/RASW

The Uses and Abuses of Humour in Social Work

Stephen Jordan

Routledge
Taylor & Francis Group

LONDON AND NEW YORK

First published 2019
by Routledge
2 Park Square, Milton Park, Abingdon, Oxon OX14 4RN

and by Routledge
52 Vanderbilt Avenue, New York, NY 10017

First issued in paperback 2020

Routledge is an imprint of the Taylor & Francis Group, an informa business

British Library Cataloguing-in-Publication Data
A catalogue record for this book is available from the British Library

Library of Congress Cataloging-in-Publication Data
A catalog record for this book has been requested

ISBN 13: 978-0-367-58238-8 (pbk)
ISBN 13: 978-1-138-47758-2 (hbk)

Typeset in Times New Roman
by Apex CoVantage, LLC

In memory of Martin Jordan 1967–2017 – *I first learnt to laugh with you, my brother*

Contents

Acknowledgements

There are many people I wish to thank for their help and support, without whom this book would not have been possible.

At Routledge I'd like to thank Claire Jarvis who set the ball rolling, Kevin Kelsey and Georgia Priestley for their encouragement and support.

Over the years, there are some great people I have had the good fortune to work with over the years Julia Blackstone, Patrick Cahillane, Ellie Hal-Fead, David Britton and many others who taught me the value of humour at work.

My colleagues in the social work team at the University of Essex, who have been supportive of my efforts, particularly Janet Phillips, Ian Harris, Sarah Wiblin, Gert Scheepers, Olivia Hanson, Andrew Thomas, Hazel Cutts, Aaron Wylie and Vasilios Ioakimidis.

There are a number of people who I worked with at the Tavistock during my studies and this book owes much to their wise and insightful advice, particularly Andrew Cooper, Clare Parkinson, Jo Finch, Judy Foster, Stephen Briggs and Bernadette Wren.

I want to thank all the practitioners who took part in my original research and all the children, families, foster carers and colleagues who shared their good humour with me over the years of practice, particularly Louis for your friendship and kind words.

And last but not least this book would not have been possible without the support and endless patience of my family Sue, Joseph and Lilian.

1 Introduction

'Spectators of our own lives'

A social worker asks a colleague: 'What time is it?'
The colleague answers: 'Sorry, I don't know. I don't have a watch.'
The first one says: 'Never mind! The main thing is that we talked about it.'

Introduction

I feel honoured to have been a social worker and often amazed of what social work can achieve when it is practiced well, despite the challenges. In an interview in which Wunmi Mosaku talked about the 2018 Channel 4 social work drama 'Kiri' she said that social work is

> a stressful job and all they're doing is trying to support where there isn't support. They go into it because they want to help, and if things go wrong it's so hard, because how can you tell if someone is going to do what you say, or they're lying? You can be cynical, or you can trust them, and that can have equally a positive or negative effect on a situation or a child's life.
>
> (Saner, 2018)

It seems to me that social work is a uniquely human endeavour to help others, which relies heavily on the skills and ability of the individual social workers, often in situations which feel like having to make the least worst decision, amongst a very limited range of ways to help. I want others to think well of the social work profession and social workers, and in this context, it might appear counterintuitive to write a book about the uses of humour in relation to social work, however the central tenet of this book is that a study of humour has much to offer, particularly in terms of understanding people, social workers and the social work profession.

Whilst social work might be considered a relatively new profession when compared with doctors, nurses or teachers, it is a profession which has struggled to be confident in its place in the world. My suggestion is that humour by exposing the flesh beneath the skin can sometimes allow us to examine the real human component of what makes social work possible. In this book I examine the times humour has been used, including when it has been used to ridicule social work. These is not an attempt to attack a profession I care deeply about, but to increase understanding of social work practice.

Mark Griffiths in his book *The Challenge of Existential Social Work Practice* posits the question, 'how do we bring about the full potential of social work in a new modern technological era?' (Griffiths, 2017, p. 14). Some authors (e.g. Vaitulionytė and Naujanienė, 2016) see humour as another creative tool for social workers to engage their service users and clients, and this book suggests that an examination of humour can help to unlock the potential of social work, as humour represents the most social of our social selves.

No book has yet been published about social work and humour, so this book makes a unique contribution to social work thought and practice. This book was born out of a desire to demonstrate my affection and love for social work in perhaps the strangest of ways, but the most collective of ways, the ways in which people laughed about social work, in the uncomfortable ways social workers laughed at others or themselves and in the way people had told funny stories about social work. In short, all the ways humour was used in relation to social work.

It is possible that humour could be seen as trivial to the very serious business of social work, and that such a book is 'frivolous' and to suggest that one might find humour or laughter in social work, could be undermining a profession which often struggles with its own self-confidence or sees itself at times as being under attack (Jones, 2012; Rogowski, 2012; Warner, 2013). This is far from my intention in writing this book and by examining humour, it is my contention that one can find enlightenment through an examination of humour.

Can you show that you value social work by finding the humour in it? The power of laughter often lies with the recipient. Humour is a worthy area of study and it's worth thinking about humour and why it works (Lockyer and Pickering, 2009). However, the process of considering humour and social work is fraught with risks. What might seem fun or humourous to some can have tragic consequences for others, and in a profession tasked with protecting vulnerable people from abuse, the use of humour is often a high wire act where things can sometimes impact negatively on people's lives. The use or abuse of humour can have tragic consequences, for example, after a hoax call to a nurse caring for the pregnant Duchess of Cornwall in December 2012, the nurse took her own life. In a tearful apology to the family at the inquest, one of the DJ's who made the hoax call said: 'I urge you to speak up . . . and consider the feelings of others when trying to make a joke' (Davies, 2014).

This book has been a struggle and I have felt in one sense to be attempting to forge a different way of thinking about social work through the prism of humour, and in this context I am acutely aware, as in the example above, that both humour and tragedy are often closely intertwined. Longo (2010) supports this assertion and has argued that humour and hatred, like comedy and tragedy, are common bedfellows and, in this way, 'exclusive and diametrical oppositions . . . reveal the multiple functions of humour in effecting a broad range of social outcomes, from bonding to excluding, controlling to resisting, all of which are linked to each other; the existence of each is necessary to render plausible the other' (Longo, 2010, p. 118). The next section outlines my pathway into social work.

My journey

When I began studying social work at Bradford University it had been famous as the place where Noel Timms was professor of social work in the 1970s. Under Timms the training for the social work degree had been firmly located in psycho-dynamic and ego psychological approaches. Timms' famous work included *The Language of Social Casework* (Timms, 1968). By the time I arrived in Bradford in 1984, there was a battle between lecturers who advocated radical anti-racist/ neo-marxist ideas in social work and lecturers more closely allied to Timms and his casework/ psychodynamic ideas. This ongoing war between what was seen as 'reactionary' casework and the opposing neo-marxists/ feminists was eloquently described by Olive Stevenson in her preface to Marion Bowers book in 2005 enti-tled *Psychoanalytic Theory for Social Work Practice*. As an undergraduate it was made clear to me that if you were not active politically in challenging the status quo you were part of the problem and lectures sometimes focussed on empower-ing service users to fight the system which oppressed them.

Of course, this is a crude parody of what was a complex and challenging time for social work, although I am tempted to ask when there was not a challenging time for social work? In this respect social work has perennially struggled with its identity and the battle between what has been called 'radical' social work and casework-based social work continues to this day. This book does not attempt to reconcile those different approaches to the practice of social work, but this was the time which gave birth to me as a social work practitioner.

Eventually I qualified in 1988 as a social worker and began working for a local authority which covered economically declining mining villages in the North of England. A year after I qualified, the Cleveland case broke in the news (Campbell, 1997) and in my newly qualified mind child abuse and specifically child sexual abuse appeared to be everywhere, alongside the cases of neglect and physical abuse. Two years later in Rochdale and in Orkney children were removed from their homes due to allegations of satanic and ritual abuse, and social work seemed to be perennially in the headlines.

The degree I had completed had radicalised and energised me to think that as a practicing social worker I could make a real difference to individual lives, along-side broader collective struggles. My degree had helped me understand that I would also be working in a hostile and challenging social, economic and political environment, with some of the most marginalised sections of society. However, the extent of poverty, abuse and hostility I experienced when I started work at times felt at times shocking and overwhelming.

Despite this I found that at work there was much humour and laughter, and sometimes I was the butt of jokes, not only from the service users, but also my colleagues. At other times I was an eager participant in office practical jokes. One of my colleagues 'Mike' was an inspirational cartoonist and frequently created cartoon works of art to entertain the team, often poking gentle fun at his colleagues, the challenges of the work or celebrating the work life of a departing colleague.

Another colleague, 'Mary' loved practical jokes and on one memorable occasion once sello-taped my sandwiches to the ceiling. Recalling such incidents sounds puerile and trite in the face of dealing with child abuse and the everyday personal tragedies social work is required to address, but these experiences remain with me to this date, and still bring a smile to my face when I recall my colleagues and experiences from this time. Humour appeared, at least on the surface, to be the mechanism for aiding work relationships and enabling myself and my colleagues not only to survive in this hostile environment, but to bond and made me feel in one sense re-energised and uplifted, ready to meet the challenges, or at the end of the day helped me to process what had happened.

Interspersed with periods in teaching I continued to practice social work up until 2014 and in work I frequently experienced shared humour with colleagues and service users. Service users would sometimes make fun of me, share jokes about their experiences and colleagues would sometimes share humourous anecdotes. In the safety of the team meetings or office, as colleagues we would share our stories and experiences with each other to create humour and to change the mood from what sometimes seemed to be the unrelenting gloom of the work.

Since 1995 *Clare in the Community* has been a regular humourous depiction of a social worker and her colleagues and in 2016 the comedy series called *Damned*, starring Jo Brand, featured a team of social workers. In this context social work and humour now appear to be, if not regular bedfellows, at least on speaking terms. In this context it feels like the right time to be writing a book which takes an in-depth look at the various ways humour and social work co-exist.

What can an analysis of humour offer social work?

Social work has a complex relationship with humour. In some respects, given the serious endeavour of social work it would appear to be the last place where one would find humour, however research and my personal experiences as a practitioner suggest otherwise. Humour has been analysed and considered in relation to other professions (Barron, 1999; Lemma, 2000). Social work literature indicates that there has been in a sporadic interest in the use of humour in social work (Siporin, 1984; Witkin, 1999; Moran and Hughes, 2006; and Gilgun and Sharma, 2011), but no book which has brought together that interest.

There are the 'in-jokes' made about social work by social workers, and the stories social workers tell each other when they want to share humorous experiences. In examining these phenomena, it is possible to think about the role social work occupies in society and examine the role humour plays in the lives of social workers and the teams they work in.

Some of the literature suggests that the use humour and jokes could provide a social function in relation to social work. Romero (2005) found that team humour contributed to 'positive mental state' and that this was associated with higher reported work effort. Uttarkar (2008) in her study of mental health teams found that humour was a way of reducing the guilt felt by staff at the advantages they had over their clients. However, Sullivan (2000) argued that not only does 'gallows humour' serve the function of self-protection, managing uncomfortable or derogatory thoughts about service users, it may place further stress on social workers who struggle earnestly with their own imperfections (Sullivan, 2000). Moran and Hughes (2006) found that humour can be both a help in moderating the effects of stress in social work and utilising humour can help others to deal with situations of extreme stress. Cooper (2008) points out theories about the causes of humour share a common thread in that they are concerned with explaining what motivates individuals to enjoy humour, rather than uncovering the social mechanisms of humour.

Stories have appeared occasionally in the popular press criticising social work practice and attacking social workers for their lack of humour. For example, in 2009 a story appeared in the Daily Mail entitled 'Social workers took away my twins after I'd joked that birth spoilt my body' (Allen, 2009). The story began with the explanation that the mother had her twin babies removed from her care by social workers after she joked 'that their caesarean birth had ruined her body.' The story went on to explain that the couple had paid out for fertility treatment and it is only further into the story that it was revealed that the 6-week premature babies were taken into care, after hospital staff warned that the first-time parents were struggling to care for them.

In relation to an analysis of humour, the comments to this article were illustrative of perceptions of social work practitioners, for example, one commentator suggested social work practitioners are like *the Gestapo* and another commentator stated that: 'This is very common behaviour from social workers who prefer to go after soft, law-abiding targets instead of confronting dangerous abusers. SS claim they're over-worked, but while they persecute this innocent family they aren't spending time protecting children who need it.'

It could be suggested that the article and subsequent comments revealed was the idea of social workers as humourless, 'child snatchers' unable to filter what appeared to be 'harmless jokes' made by 'innocent parents.' The result of which was that the parents lost their children to the 'humourless' care system. This idea of humourless bureaucrats has some history, e.g. in 1991 Punch magazine published a list of world's most loathsome human beings. Amongst the list which included Saddam Hussein, Jeremy Beadle, Gazza and Nancy Reagan, social workers were the only professional group listed. This was a troubling historical poll which made social workers 'loathsome' and worthy of ridicule. Statistically Punch readers were mainly young, male, living in the South of the UK and wealthy (Thomas, 1991), so they were not representative of the population of Britain. It was in my mind significant that a poll of largely young, southern, wealthy men, people unlikely to have any direct contact with social workers, was labelling a group of ethnically diverse, predominantly female professionals as 'loathsome human beings.'

Articles such as these can create a view of social work which generates an image in the popular mind about who social workers are. Inevitably this impacts on the work or the challenges in taking the very difficult decisions social workers face every day. If in the public mind social workers are humourless, loathsome, 'Gestapo like' characters, then social workers' decisions and actions are much more likely to be called into question. No one wants to be labelled as humourless and as Billig (2005) has pointed out to say that someone lacks a sense of humour is often seen as one of the least desirable qualities a person can have.

In addition, humour can have positive benefits, for example Morreall (2009) argues that humour can foster an open, constructive attitude to mistakes, and 'laughing at a mistake can be more beneficial than sinking into self-blame or depression' (Morreall, 2009, p. 75). Morreall (2009) argues further that humour can restore personal relationships, and that one of the most effective ways of showing people we have forgiven and forgotten behaviour which may have caused offence is by joking with them. However, in the context of social work this is not unproblematic, and one only has to look at the treatment of the social workers and manager in the wake of the tragic deaths of Peter Connelly or Victoria Climbé, to conclude that it is hard to apply Morreall's approach when reviewing the outcome of such cases.

Whilst humour may be the unifying theme of our time (Ellis, 2008) the phrase 'only joking' is used to hide uncomfortable truths or unacceptable behaviour. The famous sociologist Erving Goffman (1974) suggested that the phrase 'just joking' is one of the most commonly over-used phrases in the English language. The problem is that this from humour can also be a cover for self-deception around our *true nature*. Would it be possible to imagine social work practiced without humour? In the hands of people with malevolent intent as social beings we are averse to humour when it pokes fun at us or belittles the role we play in society and ridicules our efforts and hard work. Social work has not often been given over to self-parody, that takes an awful lot of confidence and security in the society which surrounds you. Forms of humour such as sarcasm are cruel, and cruelty is not a quality welcomed by social workers or the people they work with, indeed the business of social work must be about reducing the total sum of human misery if it is about anything at all.

Uniquely perhaps, humour, jokes and laughter seem to polarise views, as humour can alienate and exclude, but can also unify and bring pleasure almost in equal measure and for that reason raises significant questions which I feel are worthy of exploration. Could social workers be considered to be trustworthy if they were finding humour in their practice, and could such behaviour therefore be considered to be unethical? And yet given that humour, laughing and jokes are so universal and featured so much in my own practice should this be ignored or set aside as 'frivolous.' There is also a contradiction or paradox at the heart of research into humour, as many have written about the importance of humour (Sullivan, 2000; Newirth, 2006; and Cooper, 2008), but there is also a fear of taking such a frivolous subject as jokes and humour seriously. It is this complex and contradictory position which exists at the heart of this book, but as Metcalfe (2004) suggested exposing contradictions in complex social phenomena has a useful track record as a creative 'way of knowing.'

It is hardly controversial to suggest that most people like to laugh and enjoy humour, as Billig (2005) and Dessau (2012) point out there is a multi-million-pound entertainment industry given over to the purpose of making people laugh. Possessing a good sense of humour has long been viewed as a key to success in personal relationships (Guéguen, 2010; DiDonato et al., 2013). Billig (2005) has pointed out humour is at the heart of social life, 'not in the ways of easy pure creative enjoyment, but at its core humour is the no less easily admired practice of ridicule' (Billig, 2005, p. 2). So, on the one hand humour provides us with an insight into serious social, interpersonal and organisational phenomena and helps us understand the position of social work in society.

Social work is a politicised arena and is vulnerable to political changes and criticism, particularly when children or vulnerable adults die, this is due in part to the marginal position it occupies in society. Jokes by their nature enable one to obtain more than one interpretation of reality and humour can rely on the existence of numerous complex realities for its comedic effect.

I end this justification for writing this book with this quote from Critchley (2002) who pointed out that 'humour is an exemplary practice because it is a universal human activity that invites us to become philosophical spectators upon our lives' (Critchley, 2002, p. 18), and in this sense humour and jokes occupy this spectator position, an opportunity to look at a particular important aspect of my life, social work, and it is in this spectator spirit that this book is written.

Terminology

In this book I utilise the terms 'service user' and 'client' interchangeably. Although the word 'client' has largely been replaced by the word 'service user' both have similar negative connotations. More recently the term 'expert by experience' has been used, for example by my own University and whilst this feels like a more empowering term, it is not something which has yet become part of popular social work parlance. Terminology and how and why political correctness and the way social workers talk about people who are recipients of services is considered in Chapter 6.

The structure of this book

Some chapters utilise examples which are gathered as a result of the author's research, others include jokes which have been collected from the internet and written sources, or collected anecdotally by the author. Some of the quotes cited in this book first appeared in Community Care's CareSpace Forum, which was an online community hosted by Community Care magazine and closed in 2013.

Each chapter considers the implications for practice often using the words of practitioners gathered as part of the authors research and concludes with a statement about the key themes from the chapter. Chapters utilise jokes, humourous anecdotes and the words of practitioners to contextualise humour in relation to contemporary social work practice.

Laughter and humour have been the focus of study for many theorists of human behaviour and as the first two chapters (2 and 3) look at the key theories and definitions of humour and the theories of humour and the workplace, located in the range of theories and explanations about humour. Some definitions of humour including, for example: anecdotal, deadpan and droll, epigrammatic, farcical, gallows, ironic, teasing behaviour and schadenfreude are looked at in the context of social work. Inevitably this is not a comprehensive list of different types of humour but considers some the most common forms of humour illustrated in this book. Five theories of the causes of humour are considered.

Chapter 3 examines the use of humour and the workplace, given that there have been many changes to the workplace in the UK for social workers, few of which have had positive impacts on the working lives of social workers. This chapter examines the research on humour in organisations and the workplace. The relationship between stress and work is then looked at, together with the issues of emotional resilience, social work identity and the current context of social work practice in Britain.

Social work humour is starting to appear more frequently in the popular media, and in Chapter 4 examples such as 'Damned' and Clare and the Community are considered, alongside some examples form TV series such as the *Simpsons* and *South Park*. Chapter 5 considers the growth of relationship-based practice and makes a case for locating humour use as part of the relationship based practice with colleagues and service users.

Jokes and humour cause problems and have been used oppressively. Social work is a profession with explicit values to challenge discrimination and Chapters 6 and 7 consider the problematic uses of humour in terms of 'political correctness' and the anti-social aspects of humour. In the conclusion the book is ended by outlining the overall themes of this original and unique examination of social work humour.

References

Allen, V. (2009) Social workers took away my twins after I'd joked that birth spoilt my body *Daily Mail* June 20.

Barron, J. W. (1999) *Humour and Psyche: Psychoanalytic Perspectives* Hillsdale: The Analytic Press.

Billig, M. (2005) *Laughter and Ridicule towards a Social Critique of Humour* London: Sage Publications.

Bower, M. (2005) *Psychoanalytic Theory for Social Work Practice-Thinking under Fire* London: Routledge.

Campbell, B. (1997) *Unofficial Secrets: Child Abuse: The Cleveland Case* London: Virago Press.

Cooper, C. (2008) Elucidating the bonds of workplace humor: A relational process model *Human Relations* 61, 1087–1115.

Critchley, S. (2002) *On Humour: Thinking in Action* London: Routledge.

Davies, C. (2014) Prank call DJ says sorry to family of nurse after inquest *Guardian* September 13.

Dessau, B. (2012) *Beyond a Joke: Inside the Dark Minds of Stand-up Comedians* London: Arrow.

DiDonato, T. E., Bedminster, M. C. and Machel, J. J. (2013) My funny valentine: How humour styles affect romantic interest *Personal Relationships* 20(2), 374–390.

Ellis, I. (2008) *Rebels Wit Attitude* Berkely CA: Soft Skull Press.

Guéguen, N. (2010) Men's sense of humour and women's responses to courtship solicitations: An experimental field study *Psychological Reports* 107, 145–156. doi: 10.2466/07.17.PR0.107.4.145-156.

Gilgun, J. F. and Sharma, A. (2011) The uses of humour in case management with high-risk children and their families *British Journal of Social Work* (2011), 1–18.

Goffman, E. (1974) *Frame Analysis* New York: Harper.

Griffiths, M. (2017) *The Challenge of Existential Social Work Practice* London: Palgrave Macmillan.

Jones, R. (2012) The best of times, the worst of times: Social work and its moment *British Journal of Social Work* October 8. DOI: 10.1093/bjsw/bcs157

Lemma, A. (2000) *Humour on the Couch* London: Whurr.

Lockyer, S. and Pickering, M. (Eds.) (2009) *Beyond a Joke: The Limits of Humour* Basingstoke: Palgrave Macmillan.

Longo, M. (2010) Humour use and knowledge-making at the margins: Serious lessons for social work practice *Canadian Social Work Review/Revue canadienne de service social* 27(1), 113–126.

Metcalfe, M. (2004) *Creative Contradictions* Adelaide: University of South Australia (City West).

Moran, C. C. and Hughes, L. P. (2006) Coping with stress: social work students and humour *Social Work Education* 25(5), 501–517.

Morreall, J. (2009) Chapter 3 Humour and the conduct of politics from in Lockyer, S. and Pickering, M. (Eds.), *Beyond a Joke: The Limits of Humour* Basingstoke: Palgrave Macmillan.

Newirth, J. (2006) Jokes and their relation to the unconscious: Humour as a fundamental emotional experience *Psychoanalytic Dialogues* 16(5), 557–571.

Rogowski, S. (2012) Social work with children and families: Challenges and possibilities in the neo-liberal world *British Journal of Social Work* 42(5), 921–940.

Romero, E. J. (2005) The effect of humour on mental state and work effort *International Journal of Work Organisation and Emotion* 1(2).

Saner, E. (2018) Bafta-winner Wunmi Mosaku: 'I'm glad my eyes were opened after Brexit: It was an outburst of ugliness *The Guardian* www.theguardian.com/tv-and-radio/2018/jan/05/wunmi-mosaku-kiri-brexit-racism-british-nigerian-identity-channel-4-drama (accessed 14/1/18).

Siporin, M. (1984) Have you heard the one about social work humor? *Social Casework: The Journal of Contemporary Social Work* 65, 459–464.

Sullivan, E. (2000) Gallows humour in social work practice: An issue for supervision and reflexivity *Practice* 12(2), 45–54.

Timms, N. (1968) *The Language of Social Casework* London: Routledge & Kegan Paul.

Thomas, D. (1991) Punch Magazine 150 Years of Punch *Perrier Poll* July 1991.

Uttarkar, V. (2008) *An investigation into staff experiences of working in the community with hard to reach severely mentally ill people*. Thesis submitted for the award of Professional Doctorate in Social Work, University of East London in collaboration with the Tavistock Clinic October 2008.

Vaitulionytė, G. and Naujanienė, R. (2016) The uses of humor in social work practice: Analysis of social workers' experience *Social Work Experience and Methods* 18(2), 51–68.

Warner, J. (2013) Social work, class politics and risk in the moral panic over Baby P *Health, Risk & Society* [Online] 15, 217–233. http://dx.doi.org/10.1080/13698575.2013.776018

Witkin, S. L. (1999) Taking humour seriously *Social Work* 44(2), 101–104.

2 Definitions and context 'self defence against the slings and arrows of an unfriendly world'

Two social workers were walking through a rough part of the city in the evening. They heard moans and muted cries for help from a back lane. Upon investigation, they found a semi-conscious man in a pool of blood. 'Help me, I've been mugged and viciously beaten,' he pleaded. The two social workers turned and walked away. One remarked to her colleague: 'You know the person that did this really needs help.'

Laughter, happiness and humour

I start with Critchley's (2002) position is that we could consider humour 'phenomenologically,' and as a social practice that is signified by physical effects such as 'laughing, giggling, grinning and smiling, and emotional affects including joy, relief, surprise, excitement and enthusiasm' (Karlsen and Villadsen, 2015, p. 517). We can often assume that crying and tears are a response to sadness and signifies that someone is unhappy, but can we assume that when someone laughs they are happy? This is not straightforward, as we can of course cry 'tears of joy' but how do humans know something is funny and humourous unless someone else laughs at it? In another sense there can be cruel laughter, as there is cruel humour. For centuries philosophers have concluded that there is no direct correlation between the fact that just because somebody is laughing, they are happy or finding something humorous. Aristotle, Plato and Hobbes all pointed out that 'laugher is not necessarily requisite for or synonymous with humour' (Longo, 2010, p. 115). Indeed, the earliest exponents of theories of humour all viewed it as a cruel and spiteful aspect of human interactions, often involving ridicule at the perceived misfortunes of others: 'the passion of laughter is nothing else but sudden glory arising from a sudden conception of some eminency in ourselves by comparison with the infirmity of others or our own formerly' (Hobbes cited in Koestler, 1964, p. 53).

Any inspection of the internet can yield a number of tragic cases which highlight children and young people who have taken their own lives due to bullying and these are often directly related to humour and 'fun' being made of them. The sad reality is that laughter and humour made at someone else's expense can have tragic consequences. The numbers of children who turn to suicide as a result of bullying particularly via the internet appears to be increasing (The Childrens Society and Young Minds, 2018) and this led the Children Commissioner for Wales to publish specific advice over this issue (Children's Commissioner for Wales, 2017), so it obvious that the cruelty of laughter has direct negative impact. In addition, there are a number of social situations in which loud laughter would be unacceptable, not least of which would be a funeral or in response to someone disclosing abuse. Rose Coser (1959) in her famous study of laughter in hospital settings argued that laughter is a shared response, although in a world where social media is frequently the way people communicate, can you hear laughter online? Is laughter constrained by less direct human interaction? For Baudelaire (cited in Coser, 1959) the power of laughter is with the person who laughs, not the object of the humour. While laughter can represent what Nelson (2012) has termed our intimate connection with others it can also represent the most extreme forms of detachment and cruelty.

Perhaps this is best understood in the context of Duchenne and non-Duchenne laughter. 'Duchenne is the genuine, uncontrolled, involuntary expression of positive affect. Non-Duchenne laughter . . . is more of a conversational insert' (Nelson, 2012, p. 22). In this respect Duchenne laughter is involuntary and emotionally driven and can be recognised as the belly laugh or giggling fit and can be contagious (Nelson, 2012). 'Non-Duchenne laughter helps to fill in pauses, encourage conversations to keep rolling, maintain the interest and attention of a partner, disguise embarrassment, or offer an apology' (Nelson, 2012, p. 24). In the next sections I consider the things which give rise to both types of laughter.

Jokes

Jokes have been a part of human culture since at least 1900 BC (McDonald, 2010), when the first recorded joke was found in Sumerian culture. According to research by McDonald (2010) this was a joke about flatulence and is believed to be the world's oldest known joke.[1] The fart joke remains a popular source of humour today i.e. you can hear fart jokes told in any pub or school playground nowadays, suggesting that what makes people laugh has remained largely the same for several thousand years.

Jokes occupy a particular role in society and as Wittgenstein in his famous quote, eloquently put it: 'a serious and good philosophical work could be written that would consist entirely of jokes' (Malcolm, 2001). There is a danger in studying something seen as trivial as jokes and humour one could risk feeding the marginalisation of social work, but as explained in the introduction this is not the intention of this book, but jokes can provide a serious insight into human behaviour, so it is possible that jokes themselves can shed light on social works position and role in society.

Chiaro (1992) stated that analysing jokes is much like dissecting a frog, as nobody is principally interested, and the frog expires. As any professional comedian will be at pains to point out, examining a joke effectively kills it. Joking about something is also a method of disengaging from the object of the joke. It could be that the jokes about social workers reveal some of the hostility with which society regards social work or at least help people to manage their uncomfortable feelings about social work or the task of social work. Given the unpleasant and grim tasks that social workers often engage with it's hardly surprising that some of the key tasks which jokes achieve is to distance the teller and the recipient of the joke from the grimness of the work.

Palamedes, a Greek hero who outwitted Odysseus, is said to have invented the joke (Holt, 2008). The emperor Augustus was believed to have compiled no fewer than 150 joke books, but only one book of jokes survived from the earliest period of recorded history entitled *Philogelos* or lover of laughter (Holt, 2008). Jacobson (2010) argued that humour is nothing if not critical and the Greeks valued comedy higher than tragedy, as comedy 'affirmed the vigorous and unpredictable livability of life' (Jacobson, 2010).

The ancient Sumerian fart joke was found to be about 4,000 years old, although jokes are more often not the *platonic* version of an unchanging idea, and are an historical form which changes over time, often driven by the cultural context from which they emerge (Holt, 2008). Lewis (2009)argued that particular societies give rise to particular cultural forms of expression, ancient Greece had its myths and Elizabethan England had its plays. The former Soviet Union and Eastern bloc of communist countries had humour as a form of political cultural expression which challenged the ruling orthodoxy.

Freud (1960) made a distinction of types of jokes and argued that one type of joke is non-tendentious or innocent humour, and he gave the example of one: 'not only did he disbelieve in ghosts, but he was not even frightened of them' (Freud, 1960, p. 92). Freud (1960) pointed out that children engage in such jokes, which are innocent, mildly amusing, and playful and reveal no unconscious hidden agendas. Freud (1960) made a distinction between humour and jokes and pointed out that for humour to take place only one person is necessary, for comicality two persons are necessary and for a joke three are necessary: the teller, the listener and the person who it is directed at. As Howitt and Owusu-Bempah (2009) point out the joke teller and the listener have active roles in making the joke work. In summary humour can be a solitary event, but jokes require social interaction. As the example from Ancient Sumer illustrates jokes are often recorded and lend themselves to analysis. Laughter is a peculiarly human trait (Coser, 1959) and occasions for humour invite closeness in relationships and involves reciprocity.

Mik-Meyer's study suggest that jokes have a particular role to play in relation to social work e.g. 'joking is far from a neutral act. By directing a joke at a specific person, social workers can "demand" a response from that person, because of the general interaction humour that requires listeners of jokes to respond with a laugh' (Mik-Meyer, 2007, p. 23). Jokes are therefore particular social constructs which are determined by the teller and receiver.

In contrast to the occurrence of a recorded joke an occurrence of spontaneous humour, is harder to analyse, and by their nature they are seldom recorded to each other. Spontaneous and humorous anecdotes are by their nature challenging to research, as they are dependent upon being recorded, however the growth of digital recording technology has enabled more instances of spontaneous humour to be recorded. YouTube provides many examples of orchestrated occurrences of humourous encounters.

There have been publications which have collected jokes about social work, although there is a paucity of published social work jokes, for example the *Penguin Dictionary of Jokes* (2003) contains no references to social work, despite having three psychologist jokes and two jokes about sociologists (Metcalf, 2003). Young (2012) published a book entitled *The Best Ever Book of Social Worker Jokes; Lots and Lots of Jokes Specially Repurposed for You-Know-Who*, this was a book of as Young indicates 'tired, worn out jokes' applied to social work, and rather than original jokes these relied on recycled jokes which had previously been applied to other professions e.g. traffic wardens. The next section considers the different types of humour which exist.

Types of humour

Anecdotal humour

Its common to hear social workers sharing their stories. A published example of this is King and Brown (2015) who brought out a book in 2015 which featured anecdotes (some of which were humourous) based on their experiences of being social workers. For example:

> mother has requested that the children come into care. On arrival mother has lost the keys to the door. My initial point of contact with the mother is to talk through the bay window. Unfortunately, this gets us nowhere because the children begin to get upset and confused. . . . With no other option I make the decision to climb through the window. . . . The kids think it's hilarious to see the social worker climbing through the window. . . . Definitely a funny image of me I'll always remember.
>
> (King and Brown, 2015, p. 11)

Many social workers share anecdotes about their work and the reasons for this is explored in some of the chapters which follow.

Black humour

Black humour (from the French *humour noir*) is a term which originated from André Breton to describe a genre of humour which arises from cynicism and scepticism. Black humour was often a satire on the topic of death (Haynes, 2006) and seen to be popular amongst the medical profession. The purpose of black comedy is to make light of serious and often taboo subject matter, and some comedians use it as a tool for exploring vulgar issues, thus provoking discomfort and serious thought as well as amusement in their audience (Dessau, 2013).

Deadpan humour

Social workers frequently work alongside colleagues who they describe as being humourous in a serious and dry way, but a deadpan form of humour is most often associated with performance. In relation to social work such an example of this can be found in the performances of Debbie Greaves and Jim McGrath (Miller, 2016) who base their deadpan humour on observations of social work and office politics 'Comedy should hold up a mirror to life, deconstructing idiosyncrasies to get people to laugh at the system' (Miller, 2016).

Epigrammatic

Humour consisting of a witty saying such as 'Humour must not professedly teach and it must not professedly preach, but it must do both if it would live forever' – Mark Twain (Twainquotes.com, 2018). In relation to social work, one can see signs up in offices along the lines of 'you don't have to be mad to work here, but . . . ' or 'keep calm and make coffee.'

Farcical

This form of humour is based on improbable coincidences, which together with satirical elements, create moments of humour. Whilst it may be suggested that some social workers in offices see themselves as places where farce takes place, it's more common to experience this form of humour in fiction and this is covered in Chapter 4 on fictionalised versions of social work humour.

Gallows humour

Gallows humour has been viewed as jokes and humour about one's own or other people's suffering, often in the face of very serious unpleasant circumstances and frequently told by those who have shared adversity (Sullivan, 2000). For Freud (1950) such humour was the ego's opportunity to gain pleasure and has been described as 'grotesque satire' (Dessau, 2012). Unlike what has been termed 'sick' humour gallows humour is laughter at one own circumstances (Moran and Hughes, 2006) and others have described it as 'macabre or bad taste' (Sullivan, 2000).

Ironic

Whilst not all irony is humourous, Roose et al. (2012) argue that ironical forms of humour can be used by social workers to manage the ambiguity of the work, as ironic statements open a dialogue and help social workers to question the status quo. 'Irony refers to a trope used with a subtle sense of humour in the course of a perceptive and mild jest, during which a hidden, often perverted and contradictory meaning can be discovered that serves as an opportunity. An ironic perspective in social work refers to the reflexive and pragmatic embrace of ambiguity that serves as an opportunity for social work' (Roose et al., 2012, p. 1600). In this respect irony is form of humour which allows for creative thinking and playful challenges to everyday rules, by drawing attention to contradictions in the status quo. This also parallels Koestler (1964) idea that humour is 'a creative device useful in understanding complex and contradictory social realities' (cited in Longo, 2010, p. 117). This is an example:

> *My mother warned me I'd end up in a job like this if I didn't work harder at school.*

Teasing behaviour

Some commentators have defined 'teasing behaviour' as humourous actions or joking behaviours which is 'largely harmless' and constitutes a playful interaction (Davies, 1990), but many would question this, and Strawser et al. (2005), for example, found that teasing often had negative effects on children during their childhood and defined teasing as a type of bullying or peer victimisation, largely characterised by verbal taunts (e.g. about someone's appearance, performance or family). In this context it is hard to view teasing behaviour as 'harmless.'

Schadenfreude

Schadenfreude from the German has been defined as laughter, jokes and humour which creates pleasure at someone else's misfortune, and is often related to feelings of dislike or envy towards other people (Smith et al., 2009). Van Dijk et al. (2011) found that such emotions can be 'deserved' or 'undeserved,' but that *schadenfreude* is generally a socially undesirable emotion. It is harder to supply an example of such humour in a social work context as this would go against ethical principles. Examples of where social workers have used what might be called schadenfreude have sometimes resulted in the dismissal or formal warnings.

Sick humour

'Sick' humour has been the type of humour most associated with oppressive styles of humour, for example explicitly laughing at or making fun of people with disabilities. It has been linked by researchers to a divided and exploitative environment of 'us and them' culture (Moran and Massam, 1997) and can be particularly offensive. One example was the joke made by the Scottish comedian Frankie Boyle about Katie Price's disabled son, Harvey (eight years old at that time). In social work given that 'sick' humour has been associated with exploiting the vulnerable and making fun of less powerful groups in society, sick humour should not be something heard in social work officers, although comments made on line suggest otherwise:

> *Our office is very funny, sick funny, very sick funny, no incident, no stupid comment and nothing at all really gets away without the urine being extracted.*
> (Community Care, CareSpace Forum, 2011)

Culture and humour

Humour is universal to human beings and Mahadev Apte (1985) in his classic study pointed out that no 'humour free' culture has yet been found. This has been reinforced by Ojha's (2005) work. Social work is an international profession and often practices across international boundaries, but humour is an elusive and difficult topic to study between different cultures (Driessen, 1997), and as a result this book is located in Britain and the British cultural context of humour. Whilst some of the examples of jokes and humour may have their origins in other countries (primarily English-speaking ones), many examples are from social work in Britain and need to be read as such. Social work practitioners in the UK have also referred to the specificity of humour to British culture as central to the British psyche, e.g.

> *You often find those that taking the mickey out of someone else – it is almost like a British thing. In British culture it's kind of, it's almost a term of endearment, if people are poking fun out of you . . . whereas in other cultures they . . . couldn't get their head around that the fact that when someone likes you, they take the piss out of you.*
> (Jordan, 2015)

One cannot therefore ignore the cultural specificity of humour, and the findings here suggest that the aspect of humour is linked to British culture and cultural expectations, and humour is notoriously culture-specific. The 1970s saw an increasing interest in humour amongst anthropologists (Driessen, 1997).

The examples of humour in this book primarily utilise British culture, but also acknowledges that in places American and other English-speaking cultures have provided examples of humour in a social work context.

Theories of humour

Most of the literature on humour refers to three main explanations or theories as to why we might find something comedic, funny or humourous: 1) superiority, 2) incongruity or 3) release (also known as relief), although I would also add there are two other areas of theories, first which link humour to development and second to subversion and challenges to authority, so these are also included in this section. It's common to see in the literature on humour all three of these theories coming together to make sense of humour, e.g. Gilgun and Sharma's (2011) study found social workers in their study used humour to regulate anxiety, frustration and shock, and found largely positive aspects of humour use, including emotion regulation, and creative problem solving. Social workers in their study often used humour to express liking of service users, and this fitted with the three major theories of humour: superiority, relief and incongruity (Gilgun and Sharma, 2011). This section examines the different theories of humour in more detail.

Superiority theories of humour

Anthropological studies emphasise the commonalities across cultures of humour, the cultural and social factors in explaining the causes of humour, e.g. the inappropriateness of humour at a funeral (Lemma, 2000). Lemma (2000) has provided several examples of different cultural contexts, which in other cultural contexts would be unfunny, for example not appreciating a joke in Trinidad is equated with 'not belonging' and most cultures considered humour and laughter to be a social lubricant or even a sign of social acceptability (Lemma, 2000).

Anthropological studies illustrate the richness of humourous expressions across the globe. Carden (2003) described the findings of humour across a range of cultures including North American Indians, the Maoris, native Canadians, African-Americans, Jews, Polish and Irish and that humour is a vehicle of 'maximum consciousness' through which minority cultures reflected on their own anger and distress, particularly in the face of oppression. Carden (2003) argued that humour helped cultures preserve a sense of identity. The strategies, in the eyes of anthropologists, not only favoured in-group cohesion, but also have systematically served to question the legitimacy of exploitation and oppression. For example, Scheper-Hughes (1993) described how the impoverished people of a shantytown in northeastern Brazil used gallows humour to survive. Billig (2005a) found that many anthropological studies (Watson-Gegeo and Gegeo, 1986; Eisenberg, 1986; Miller, 1986; Clancy, 1986 – all cited in Billig, 2005a) revealed that mothers often used teasing to manage children's inappropriate behaviour. This form of social control was often used as an alternative to physical chastisement, and to establish parental authority.

Other theorists and philosophers e.g., Plato, Hobbes and Bergson (Billig, 2005a; Holt, 2008) believed that at its root all humour is interpersonal mockery and derision, and this has been referred to as 'the superiority school of thought.' Holt (2008) suggested that all laughter is a 'slightly spiritualized snarl' (Holt, 2008, p. 81). Sullivan (2000) argued that gallows humour was humour used by social workers to emphasise the difference between them and service users in terms of status, saneness, intelligence and knowledge and fits the superiority theory of humour.

Psychologists have been more interested in the mechanism of jokes and found that what makes people laugh hardest is the speed with which they get the punchline (Lewis, 2009). Whilst psychologists have focussed on the internal processes which influence the formation of humour and jokes, sociologists have tended to focus on what function jokes and humour achieve in society.

Comprehension-elaboration theory argues that the degree to which someone will enjoy humour is determined by how difficult the humour is to comprehend and by the cognitive elaboration performed after comprehending the humour (Cooper, 2008). This theory suggests that the degree to which one person will find a comment or behaviour as humourous depends on the motivation of the person conveying the joke, whether the humour is socially appropriate to their situation and whether the humour is offensive to themselves or other groups. The immediate enjoyment of the humour may decrease if the individual finds the humour to be hurtful or concludes the person expressing the humour had an undesirable motive Cooper (2008).

When discussing the cultural aspect of jokes told in the totalitarian regimes Lewis found that jokes 'revealed peoples states of mind and . . . jokes gave them courage' (Lewis, 2009, p. 5). Lewis (2009) argued that jokes were so popular in the Soviet Eastern bloc countries because they enabled people to defend against the pain of their everyday lives, resist the oppressive nature of the regimes and made themselves feel superior to their corrupt and incompetent leaders.

Humour mirrors and often expresses the moral, cultural and political themes of the age it arises in and reflects the complex dialectic of discipline and rebellion (Billig, 2005a). For example in the Middle Ages it would have been socially acceptable to mock the physically afflicted or people with learning disabilities. In the hands of an oppressed group, humour can be used to challenge authority, but used by the powerful it can be used to oppress minorities. Holmes (2000) and Mik-Meyer (2007) studied the use of joking in the workplace and identified hierarchical patterns of humour, with humour flowing downwards, as 'superiors' make jokes to 'inferiors,' or social workers to service users, rather than vice versa. Robert and Wilbanks (2012) indicate that when humour is used by high-power supervisors in an aggressive manner to control or dominate others it reinforces hierarchical difference and power, whereas cycles of reciprocated humour can reduce obvious hierarchical differences.

Billig (2005a) placed the theorists of humour within their social context, for example Hobbes' argument that as laughter reflects the sort of base, selfish motives that need to be disciplined, humans need external controls to prevent their selfishly destructive urges from running riot. Billig (2005a) suggests that a society filled with laughter would not be a happy place, but rather a place where human beings were baited unmercifully, however it also feels desirable to me to see humour as a vehicle for rebellion and subversiveness, where the powerful are baited by the powerless.

However, there is criticism of the superiority model of humour for oversimplifying humour to a binary version of power:

> The devaluation of one person (the object) by another (the subject) is the requisite condition upon which all humour emerges to convey the ostensibly superior moral views of the object. A power-over, winner-loser model of humour production such as this . . . overly simplified and absolute assessment of the binary nature of power and control depends on a reductionist view of

individuals and human experience that categorizes individuals as exclusively occupying one of two camps: the oppressor or the oppressed.

(Longo, 2010, p. 116)

The next theory attempted to avoid such binary reduction.

Incongruity theories of humour

Locke, Pascal, Kant and Schopenhauer subscribed to this body of theory and argued that humour arose from a conflict between the logical and respectable, dissolved into the illogical and absurd (Paulos, 2000; Holt, 2008). Incongruity theory is also supported in the writings of Kant and Kierkegaard (Cooper, 2008), and Wittgenstein was concerned with the humour found in nonsense, logical confusion and language puzzles (Paulos, 2000). Wittgenstein argued that something was funny because it was a logical contradiction (Paulos, 2000). Fry (1968) demonstrated that when someone tells a joke there is normally a behavioural cue that what someone is about to deliver is false, and not an everyday ordinary interaction. It is as if someone is saying I am about to tell you an incongruity here and the test is whether you will be able to understand it.

'Incongruity theory charges that humour erupts from a sudden disruption in the natural order' (Longo, 2010, p. 116) and Holt (2008) argued that when logic goes astray, laughter serves to draw attention to the fallacy and this is like Nietzsche's argument that laughter is a cure for aberrations of pure reason (Holt, 2008). Following the developmental theories of Piaget, McGhee (1971) found that children's level of cognitive functioning and their comprehension and appreciation of humour was based on violation of cognitive expectancies, and disruption of what they normally expect.

I remembered the times when my own children laughed and as a parent there are few things more enjoyable than being able to see your children laugh. Young children are more likely to find something humourous if it was incongruous to their expectations, this is an example:

> One day a man with a dog walks into a cinema. 'I'm afraid I can't let your dog in here, sir,' the manager says. 'Oh, I assure you, he's very well behaved,' the man says. 'All right then,' the manager says. 'If you're sure . . . .' After the movie, the manager says to the man, 'I'm very surprised! Your dog was well behaved, and he even seemed to enjoy the movie!' 'Yes, I was surprised, too,' says the man. 'He hated the book.'

The distinctive aspect of the incongruity theory of humour is that it does not propose an emotion behind our enjoyment of the ludicrous, and modern-day theories of humour approach this topic from an analysis of jokes (Billig, 2005a), suggesting that most jokes are based at least on one level on incongruity. I wonder that as adults we might envy a child's ease of laughter and pleasure at the ludicrous.

Humour as a psychosocial mechanism for managing emotions (release or relief theory)

This group of theorists is linked to the emotional context of humour and jokes and was later associated with Freud's ideas about the causes of humour. Humour is seen as the release of inhibitions, and the need for laughter is a process by which people vent their repressed tensions and advocated by nineteenth-century theorists such as Bain and Spencer, it has been referred to as Victorian relief theory (Billig, 2005a).

Freud's work (1960) gave birth to an extensive psychoanalytic literature on jokes and humour. Freud's key argument is that jokes, like dreams, unmask hidden truths and thereby offer a release for the mind (Lewis, 2009). When someone says of someone else they are 'a bit of a joke' this is almost always something nastier or vicious. Freud (1960) observed that the first aspect of humour begins with the infant smile. In this respect humour is one of the most primitive and fundamental aspect of our psyche and an indestructible part of our unconscious desires. Holt (2008) found that the journalist Gershon Legman collected dirty jokes and found that dirty jokes revealed the 'infinite aggressions' of men against women, supporting Freud's hypothesis. For Legman the telling of a dirty joke was equal to verbal rape (Holt, 2008).

Freud (1960) argued that jokes and humour need an audience, as jokes are primarily social affairs and the satisfaction from them comes primarily from the fact that the anxiety expressed in the joke is acknowledged and shared by the audience listening to it. In this sense the audience becomes the container for the jokes. Here the laughter of the audience confirms for the teller that the pleasure of the joke outweighs the pain and anxiety and that both the teller and the audience have survived the attack from their shared unconscious. Billig (2005a) argued that both Hobbes and Freud saw the fundamental conflict between individual desire and social order – humans are selfish, but we need to live socially, so humour allows us to manage the difficult feelings we have towards others, such as racism, hatred or the desires to hurt others. Freud (1960) argued that repression is the way through which unruly human nature is socially disciplined, and for Freud (1960) a way of understanding this as *jokes provided sense in nonsense*.

Freud (1960) argued that we repress our hostile impulses against our fellow men and make our enemies small and inferior, and in this way overcome them. Jokes then are a rebellion against authority and can be used to 'alienate our enemies' (Freud, 1960). Jones argued that 'humour is one of the chief means of self defence against the slings and arrows of an unfriendly world' (Jones, 1948 p.374), but Lemma (2000) points out that humour is more likely when one party retaliates for some provocation, than just establishing superiority over someone.

Berger (1997) pointed out that aggressive jokes break the taboo on aggressive acts, and become a mechanism for individuals, social groups or society to express its frustration without physically attacking. Berger (1997) argued that we want to believe the best of ourselves, and if we laugh at cruel or obscene jokes it is because humour and jokes are an escape from reality, or as Freud (1960) suggested an escape from the reality of ourselves. Billig (2005a) pointed out that Freud's work revealed the unpleasant and harmful nature of humour, and avoided the supposition that humour is necessarily to be applauded for being witty or clever, for example racists do not become any less racist by telling a joke (Billig, 2005b).

Barron (1999) argued that humour is more than just a manic defence, and at its most sublime is produced in the face of death. Viewed this way humour and jokes become 'safety valves' for hostilities and discontent ordinarily suppressed by individuals or groups (Coser, 1959). This is supported by research in the field, as Sullivan (2000) found that social workers in children services found the humour allowed them to vent their feelings about children, so that they could face them again. The more the social workers spoke negatively about a child, the more they 'appeared to be able to see them positively' (Sullivan, 2000, p. 48). For Longo, 2010 'Relief theory helps to elucidate the role of humour to convey experiences of marginalization and oppression and as a disarming yet acceptable tool for the expression of social resistance in informal and formal institutionalized contexts' (Longo, 2010, p. 115).

Freud's theory suggests that the pleasure we derive from jokes stems from the psychic energy used to inhibit aggressive and sexual impulses. It follows that the people who laugh hardest at malicious jokes are the ones who have most deeply buried their aggressive tendencies, but research by Eysenck suggested that those who laugh most at lewd sexual jokes are people who are least likely to be sexually repressed (Holt, 2008). This body of theory is considered in more detail in Chapter 6.

Developmental theories of humour

The core argument of this group of theorists is that humour is essential to human growth and development (Bateson, 1953; Bowlby, 1999). Some evolutionary scientists suggest that humour has played a vital role in the development of the unique intellectual and perceptual abilities of humans (Clarke, 2008). Clarke (2008) argued that when the brain finds something amusing it recognises a pattern that surprises it. On an evolutionary level this ability to recognise patterns unconsciously is an asset and the benefits of humour encouraged human beings to develop their adaptability to new circumstances (Clarke, 2008). Bateson (1953) argued that humour is an evolutionary step in the human species. He found that the ability to discriminate between messages encoded at different levels of abstraction was inherent in the development of playful activities, humour amongst them. Billig (2005a) argued that the human capacity to smile and laugh is biologically inherited, which gives infants an evolutionary advantage in how to communicate, as it provides survival factors. Robert and Wilbanks (2012) suggest that humans might be 'hardwired' to experience cycles of humour, affect and laughter, and that the positive affect in an audience that is induced by a humour creator might 'bounce back' to the creator through automatic responses to audience laughter.

Other developmental theorists have focussed on language development in children and humour. Word play is an important part of language development in children, and many researchers have highlighted how developing children enjoy jokes and riddles (King-DeBaun, 1997; Musselwhite and Burkhart, 2002).

Bowlby (1999) argued that smiling was crucial to developing attachment and Nelson (2012) has argued that laughter effect arousal and regulation from the caregiver to the baby. Nelson (2012) argued that laughter is a process for the caregiver and babies to attune to each other, and is the beginning stage of how we learn to interact with one another. Laughter and humour therefore provide a secure base for exploring the world. For Bowlby (1999) a smile can ensure responsiveness to the infant from the person providing care, ensuring physical proximity and a loving relationship. Humour therefore plays a central role in attachment, long seen as fundamental to a child's development (Bowlby, 1999; Howe, 2005, 2011). These points are reinforced by Spitz (1965), who suggested that the early stages of development a smile is motivated by the sight of the human face. Aimard (1988) found that humour is a regulator that helps to clarify and preserve the balance between the child and their family, and children use humour to negotiate and manage difficult situations.

Bollas (1995) suggested that a sense of humour is essential to human survival, and the mother who develops their baby's sense of humour is assisting him to detach from mere existence and as an adult, they will find humour in the most awful circumstances, ultimately benefiting from the origins of the comic sense. This had resonance for me in relation to social work.

Finally drawing on Chomsky's notion of 'linguistic competence' Raskin (1985) developed the theory of humour competence, which I have included in this group of developmental models of humour theories. Raskin's theoretical model focusses on the reason why one person might find something amusing whereas another person may not, and Raskin (1985) suggests is best understood by examining people's social positions, as individuals share humour with others they consider similar or 'like-minded'.

Social 'subversiveness' theories (a sub-category of superiority)

For thousands of years there have been debates over humour – the ancient Greek philosopher Plato saw humour as subversive, whereas Aristotle saw it as cathartic and educative (Ritchie, 2010). This subversive view of humour has been shared by twentieth-century writers such as George Orwell who said that 'every joke is a tiny revolution' (Lewis, 2009, p. 19). Lewis argued that jokes are powerful tools in creating change, as they are linked to resistance and challenging oppression. A group process occurs when a joke is told, as Blau noted jokes were almost always shared with a group whereas complaints were told to individuals (Blau, 1955 cited in Coser, 1959). Jokes and humour aimed at making fun of the oppressively powerful appears 'justified' and the jokes which punctured the pomposity of the Soviet Union worked to eventually undermine the social authority of the regime (Lewis, 2009). Billig (2005a) argued that joking relationships are not presumed to be necessary for the continuation of social life in general, but humour can ease the exercise of power.

Employees who are adept at using humour can adopt the role of 'sage fool' as a way of managing and expressing dissenting opinions, feeding these back to management in a less challenging way to authority (Cooper, 2008). The workplace use of humour is explored in more detail in the next chapter.

Bergson said the 'humourist is a disguised moralist' (Coser, 1959). Holmes (2000) found that humour could be a means by which subordinates could challenge power structures and make what might be thought of as 'risky statements,' in a lighthearted way. Christie (1994) argued that humour enables us to tolerate antithetical ideas. Gilgun and Sharma (2011) argued that there is an interplay between the audience, target of the joke and the joke teller, and their relationship provides the joke teller with the authority to get away with the joke (Gilgun and Sharma, 2011).

In the 1960s Schmulowitz, an assiduous collector of jokes, found that jokes contained great power (Holt, 2008). He referred to them as the 'small change of history,' and how jokes have 'detected and exposed the imposter and have saved man from the oppression of false leaders' (Holt, 2008, p. 40). Carden (2003) found that humour is developed and strategically used among physically or ideologically oppressed communities and at a socio-cultural level, to challenge their oppression. As such humour and joking behaviour appears to reflect ideological struggles between dominant and subdominant traditions. The subversive aspect of humour is considered in more detail in Chapter 7.

Conclusion

In my work as a social worker there were lots of instances when colleagues came back into the office to share their humourous anecdotes. On one occasion Julia came back from a visit where she had been asked to assess a young man with a history of mental health issues. She knocked on the door of a semi in the middle of the housing estate where most families were familiar with the social work team. As she stood waiting for the door to be answered a dog came up, sat beside her and looked up at her. The dog seemed friendly enough and she took this as a good sign for her assessment. The young man answered the door and after Julia had introduced herself, showed her in to the living room. The dog followed her in and sat beside her, as Julia perched on the settee, conducted her assessment. About half way through the assessment, the dog walked over to the armchair where the young man was seated, cocked its leg and began urinating. The young man looked at Julia and Julia smiled at the young man, and thought: 'he has different standards from me, I won't comment on it at this visit, but ask to him about hygiene on the next visit.' The young man said nothing. Bringing the interview to a close ten minutes later Julia stood up and went to the door. Just as she was about to leave the young man asked her plaintively 'Aren't you going to take your dog with you?'

This account could be analysed in terms of its relationship to incongruity, but it is also an example of anecdotal humour which exemplifies our emotional release and we could even see elements of superiority theory. Why should a profession concerned with understanding and helping people be concerned with an activity such as humour? It seems that humour is one of our most social activities, and the concern of many philosophers, anthropologists and experts of human behaviour for millennia. It's impossible not to see humour as being part and parcel of almost everything human beings do and for this reason alone impossible not to consider in relation to social work practice.

As the claim to having a sense of humour appears to be universal many researchers and commentators tend to be obliged to make universal claims about their ideas and findings (Willis, 2009). This chapter illustrates that there is no general theory of humour and instead suggests that in order to understand humour we have to draw on the myriad of historical and philosophical positions, which have often located explanations of humour in specific cultural and economic times.

Note

1 The joke is; '*Something which has never occurred since time immemorial; a young woman did not fart in her husband's lap.*'

References

Aimard, P. (1988) *The Babies of Humour* Wavre: Mardaga.

Apte, M. L. (1985) *Humor and Laughter: An Anthropological Approach* (Paperback ed.) Ithaca, NY: Cornell University Press.

Barron, J. W. (1999) *Humour and Psyche: Psychoanalytic Perspectives* Hillsdale NJ: The Analytic Press.

Bateson, G. (1953) The position of humour in human communication in Foerster, H. (Ed.), *Cybernetics* New York: Josiah Macy, Jr. Foundation.

Berger, P. (1997) *Redeeming Laughter: The Comic Dimension of Human Experience* New York: Walter de Gruyter.

Billig, M. (2005a) *Laughter and Ridicule towards a Social Critique of Humour* London: Sage Publications.

Billig, M. (2005b) Violent racist jokes: An analysis of extreme racist humour in Lockyer, S. and Pickering, M. (Eds.), *Beyond a Joke* Basingstoke: Palgrave Macmillan.

Bollas, C. (1995) *The Work of Unconscious Experience* London: Routledge.

Bowlby, J. (1999) *Attachment and Loss: Vol* I, 2nd Ed. New York: Basic Books.

Carden, I. (2003) On humour and pathology: The role of paradox and absurdity for ideological survival. *Anthropology & Medicine* 10(1), 115–142.

Chiaro, D. (1992) *The Language of Jokes Analysing Verbal Play* London: Routledge.

Christie, G. L. (1994) Some psychoanalytic aspects of humour *The International Journal of Psychoanalysis* 75, 479–489.

Children's Commissioner for Wales (2017) *Sams Story: Listening to Children and Young People's Experiences of Bullying in Wales* Published July 2017 www.childcomwales.org.uk/wp-content/uploads/2017/07/Sams-Story.pdf

The Children's Society and Young Minds (2018) *Safety Net: Cyberbullying's Impact on Young People's Mental Health Inquiry Report* www.childrenssociety.org.uk/sites/default/files/social-media-cyberbullying-inquiry-full-report_0.pdf

Clarke, A. (2008) The science of laughter *The Times* 13 September 2008.

Community Care (2011) *Care Space Forum* http://www.communitycare.co.uk/join-social-work-online-community/ (Forum closed in 2013).

Cooper, C. (2008) Elucidating the bonds of workplace humor: A relational process model *Human Relations* 61, 1087–1115.

Coser, R. L. (1959) Some social functions of laughter: A study of humour in a hospital *Human Relations* 12, 171–182.

Critchley, S. (2002) *On Humour: Thinking in Action* London: Routledge.

Davies, C. (1990) *Ethnic Humour around the World* Bloomington, IN: Indiana University Press.

Dessau, B. (2012) *Beyond a Joke: Inside the Dark Minds of Stand-Up Comedians* London: Arrow.

Dessau, B. (2013) *News: Jo Brand Defends Lee Mack News* October 10 www.beyondthejoke.co.uk/content/news-jo-brand-defends-lee-mack

Driessen, H. (1997) Chapter 12 Humour, laughter and the field: Reflections form anthropology in Bremmer, J. and Roodenburg, H. (Eds.), *A Cultural History of Humour: From Antiquity to the Present-Day* Cambridge: Polity Press.

Freud, S. (1960) *Jokes and Their Relation to the Unconscious* London: Routledge & Kegan Paul.

Freud, S. (1950) *Der Humour* www.scribd.com/doc/34515345/Sigmund-Freud-Humor-1927

Fry, W. F. (1968) *Sweet Madness A Study of Humour* Palo Alto: California First Pacific Books Paperbounds.

Gilgun, J. F. and Sharma, A. (2011) The uses of humour in case management with high-risk children and their families *British Journal of Social Work* 42(3), 1–18.

Haynes, D. (2006) The persistence of irony: Interfering with surrealist black humour Textual *Practice* 20(1), 25–47. DOI: 10.1080/09502360600559761.

Holmes, J. (2000) Politeness, power and provocation: How humour functions in the workplace *Discourse Studies* 2, 159–185.

Holt, J. (2008) *Stop Me If You've Heard This Before: A History and Philosophy of Jokes* London: Profile Books.

Howe, D. (2005) *Child Abuse and Neglect: Attachment, Development and Intervention* Basingstoke: Palgrave Macmillan.

Howe, D. (2011) *Attachment Across the Lifecourse: A Brief Introduction* Basingstoke: Palgrave Macmillan.

Howitt, D. and Owusu-Bempah, K. (2009) Chapter 2 Race and Ethnicity in Popular Humour in Lockyer, S. and Pickering, M. (Eds.), *Beyond a Joke: The Limits of Humour* Basingstoke: Palgrave Macmillan.

Jacobson, H. (2010) The Finkler Question: Howards Jacobson views of humour *The Guardian*, Saturday 9 October 2010.

Jones, E (1948) *Papers on Psycho-Analysis*. London: Balliere Tindall & Cox. Revised and enlarged editions, 1918, 1923, 1938, 1948 (5th edition).

Jordan, S. (2015) *That joke isn't funny anymore: Humour, jokes and their relationship to social work*. Professional doctorate thesis, University of East London.

Karlsen, M. P. and Villadsen, K. (2015) Laughing for real? Humour, management power and subversion *Ephemera* 15(3), 513–535. ISSN 1473-2866 (Online) ISSN 2052-1499 (Print) www.ephemerajournal.org

King, S. and Brown, N. (2015) *Brummie Girls Do Social Work* Oxford: Shrewsbury You-Caxton Publishers.

King-DeBaun, P. (1997) *Computer fun and adapted play: Strategies for cognitively or chronologically young children with disabilities part 1 and 2*. Proceedings of Technology and Persons with Disabilities Conference, California State University, Northridge, CA., USA.

Koestler, A. (1964). *The Act of Creation*. New York: Penguin Books.

Lemma, A. (2000) *Humour on the Couch* London: Whurr.

Lewis, B. (2009) *Hammer and Tickle* London: Orion Books.

Longo, M. (2010) Humour use and knowledge-making at the margins: Serious lessons for social work practice *Canadian Social Work Review/Revue canadienne de service social* 27(1), 113–126.

Malcolm, N. (2001) *Ludwig Wittgenstein: A Memoir* Oxford: Oxford University Press.

McDonald, P. (2010) Paul McDonald: Heard the one about the oldest joke in the world? It's a cracker! *The Independent* (accessed 19/12/2010).

McGhee, P. (1971) Cognitive development and children's comprehension of humour *Child Development* 42(1), 123–138.

Metcalf, F. (2003) *Penguin Dictionary of Jokes* London: Penguin.

Mik-Meyer, N. (2007) Interpersonal relations or jokes of social structure? Laughter in social work *Qualitative Social Work* 6(9).

Miller, N. (2016) Interview social workers do standup: 'A good laugh makes you feel better' *The Guardian* Tuesday November 1.

Moran, C. C. and Hughes, L. P. (2006) Coping with stress: Social work students and humour *Social Work Education* 25(5), August, 501–517.

Moran, C. C. and Massam, M. (1997) An evaluation of humour in emergency work *The Australian Journal of Disaster and Trauma Studies* (3), 36–42.

Musselwhite, C. and Burkhart, J. (2002) *Social scripts: Co-planned sequenced scripts for AAC users*. Proceedings of Technology and Persons with Disabilities Conference, California State University, CA, USA www.csun.edu/cod/conf/2002/proceedings/csun02.htm

Nelson, J. K. (2012) *What Made Freud Laugh: An Attachment Perspective on Laughter* London: Routledge.

Ojha, A. (2005) Jablin's organizational assimilation theory and humor: A closer look at the ontological and epistemological issues of how humor can be used to assimilate into an organization *Journal of Organizational Culture, Communications and Conflict* 9(2), 131.

Paulos, J. A. (2000) *I Think, Therefore I Laugh: The Flip Side of Philosophy* London: Penguin.

Raskin, V. (1985) *Semantic Mechanisms of Humour* Dordrecht: D Reidel.

Ritchie, C. (2010) Against comedy *Comedy Studies* 1(2), 159–168.

Robert, C. and Wilbanks, J. E. (2012) The wheel model of humor: Humour events and affect in organizations *Human Relations* 65. DOI: 10.1177/0018726711433133

Roose, R., Roets, G. and Boueverne-De Bie, M. (2012) Irony and social work: In search of the happy sisyphus *British Journal of Social Work* 42, 1592–1607.

Scheper-Hughes, N. (1993) *Death without Weeping: The Violence of Everyday Life in Brazil* Berkeley: University of California Press.

Smith, R. H., Powell, C. A. J., Combs, D. A. and Schurtz, D. R. (2009) Exploring the when and why of Schadenfreude *Social and Personality Psychology Compass* 3(4), 530–546.

Spitz R. A. (1965) *First Year of Life: A Psychoanalytic Study of Normal and Deviant Development of Object Relations*. Madison, Connecticut: International Universities Pr Inc.

Strawser, M. S., Storch, E. A., and Roberti J. W. (2005) The teasing questionnaire—revised: Measurement of childhood teasing in adults *Journal of Anxiety Disorders* 19(7), 780–792.

Sullivan, E. (2000) Gallows humour in social work practice: An issue for supervision and reflexivity *Practice* 12(2), 45–54.

Twainquotes.com (2018) Originally from *Mark Twain in Eruption* www.twainquotes.com/Humor.html (accessed 3/3/18).

Willis, K (2009) Chapter 6 Merry Hell: Humour competence and social incompetence in Lockyer, S. and Pickering, M. (Eds.), (2009) *Beyond a Joke: The Limits of Humour* Basingstoke: Palgrave Macmillan.

Van Dijk, W. W., Ouwerkerk, J. W., Wesseling, Y. M. and Van Koningsbruggen, GM (2011) Towards understanding pleasure at the misfortunes of others: The impact of self-evaluation threat on schadenfreude *Cognition & Emotion* 25(2). DOI: 10.1080/02699931.2010.487365 pages 360-368

Young, M. G. (2012) *The Best Ever Book of Social Worker Jokes: Lots and Lots of Jokes Specially Repurposed for You-Know-Who* New York: Dolyttle and Seamore.

3 Humour and the workplace
The bonds of stress

The social worker asked the pub landlord: 'What's the difference between your job and mine?' The landlord replied, 'I did not have to go to University, I learned to mix a little of this with a little of that on the job and yet people will tell me their innermost thoughts, while you went to University for 3 years, paid thousands and thousands of pounds, sit session after session using technique after technique, and they may still not talk to you.'

Introduction

Social work has had an extensive and complicated history in the UK, and its origin has been located in the philanthropy of Victorian charity organisations in the nineteenth century. The history and role of social work in the UK has been debated and analysed extensively before (for example Seed, 1973; Jordan, 1984; Payne, 1996; Mullaly, 1997; Lymbery, 2005; Thompson, 2005; and Horner, 2007). Payne (1996) famously suggested that social work could be reduced to three perspectives: 'individualist reformist,' 'reflexive-therapeutic' and 'socialist-collectivist.' Such perspectives reflect the history of social work in the UK as it developed from charitable work and good 'deeds,' through to becoming part of the state and on to an international movement concerned with social justice and collective action.

One could also add a fourth perspective – 'managerialist-technicist' (Harlow, 2003), which reflects the increasing role of the private sector in service delivery, a preoccupation with targets (as measured by Ofsted) and 'value for money.' The increasing role of the private sector has been linked to the rise of neoliberal political agenda for social work under the New Labour and Conservative/Liberal governments (Ferguson, 2008; Harlow et al., 2013).

The choice of which definition is contingent on the historical context, practitioner's personal skills or knowledge, agency requirements or the needs of the service user. Jordan (1984) suggested that arguments about what social work practice is, are really arguments about the causes of social problems and the solutions to those problems (Jordan, 1984). Mullaly (1997) argued that there are essentially two perspectives of social work practice, first the 'conventional' perspective where professionals engage in statutory activities focusing on individual service users, where practitioners work mainly in an 'individualist-reformist' and 'reflexive-therapeutic' way. Mullaly (1997) contrasts this with the 'progressive' perspective, which informs a smaller number of the profession who consider individuals' problems the result of an inequitable society. Social work practice in this context has focused on challenging societal inequalities, such as: sexism, poverty, disablism and racism. The International Federation of Social Workers has defined social work as a profession which 'promotes social change, problem solving in human relationships and the empowerment and liberation of people to enhance well-being' (Hare, 2004) and in the end 'social work is a contingent activity, conditioned by and dependent upon the context from which it emerges and in which it engages' (Harris, 2008).

The current context of social work in the UK is a largely public sector profession which has suffered both a decline in funding as the current government continues with austerity and a decline in public trust, for example in 2012 Hope-Hailey et al. suggested that trust in the public sector and public confidence in social work organisation in particular was at 'an all-time low' based on longitudinal surveys, such as the Edelman's Trust Barometer (Hope-Hailey et al., 2012). The death of Peter Connelly (Baby P) in 2007 saw a dramatic increase in the number of care applications from local authorities, which suggest that negative media reporting has a direct influence on social work practice and this too can contribute to a climate of fear and blame. 'Recent high-profile scandals, such as the damning reports into social work practices in Rochdale, Oxford and Rotherham serve to undermine public trust further, and raise questions about the ethical and moral conduct of our social service institutions and their employees' (Legood et al., 2016, p. 1874). It could be argued that increased media attention and heightened public awareness of social workplaces a strain on those working within the profession.

Rogowski (2012) has pointed out that decades of de-professionalisation, managerialism and the move to a technical-rationalist and business model has dominated the discourse around social work (Rogowski, 2012). Rogowski (2012) has further argued that social workers are often so focussed on 'getting the job' done that they are in danger of losing sight of what and who they are, including their professional uniqueness and style of intervention. In 2008, the government commissioned the Social Work Task Force, which recommended the establishment of the College of Social Work, alongside 15 recommendations for improving and reforming social work.

However the recommendations of the Social Work Task Force reforms, including the establishment of the College of Social Work in 2012, did not last long in raising the profile and the College closed after only three years and the General Social Care Council was replaced by the Health Care Professions Council as the body which regulated social work, which also emphasises that social workers must practice lawfully, safely and effectively (HCPC, 2017).

Social work practice in Britain therefore exists in a repressive atmosphere, in which many politicians, members of the public and the media appear, if not explicitly hostile to social work, often to display little confidence in the profession, particularly in the wake of child deaths such as Daniel Pelka and Keanu

Williams (Bennett, 2013; Elkes, 2013). Freud's belief that every society requires a certain level of repression to function appears to resonate in this current repressive climate (Billig, 2005). Studies of joke telling in Eastern Bloc countries (Lewis, 2009) and anthropological studies of oppressed groups use of humour (Carden, 2003; Scheper-Hughes, 1993) suggest that jokes, joking behaviour and humour flourish under repression, as people seek ways to express themselves through humour. Oppressed groups can rebel and challenge oppression though 'safe mechanisms' such as humour.

For Warner (2017) it is the state that provides the mandate for social workers to intervene in private family life. The idea that such intervention is right in *certain circumstances* to protect a child continues to achieve a broad social consensus. Therefore, despite social work's apparent abject failings in the eyes of the media and others, social work as a profession survives, albeit with endless reform and reinvention. For this reason the next section looks at the relationship between resilience and humour.

Emotional resilience and humour

There has been increased interest in the relationship between social work and resilience (Kinman and Grant, 2011). The word resilience was first applied in the mid-seventeenth century in the context of the properties of metals and whether a piece of metal could go back to its original shape, and originated from Latin as the verb 'resilire' meant 'leaping back.' Boris Cyrulnik, a Jewish psychiatrist and neurologist, popularised the concept of resilience, as he survived the Nazis while his parents were murdered. 'Resilience is a complex and multi-faceted construct, referring to a person's capacity to handle environmental difficulties, demands and high pressure without experiencing negative effects' (Kinman and Grant, 2011, p. 262). Kinman and Grant (2011) suggest that resilience can help individuals manage the negative effects of work-related stress and that as a result resilience is an important aspect of a social worker's working life and workplace environment. The key point is that a protective factor is humour use by a resilient individual.

The 1980s witnessed a burgeoning interest in the investigation of protective factors in children who were coming into the care system. Why did some children appear to do well in care, whilst others rapidly declined? It was thought that protective factors including individual and environmental characteristics moderated the negative effects of stress and resulted in more positive behavioural and psychological outcomes in at-risk children than would have been possible (Masten and Garmezy, 1985). Children whose behaviour reflects these protective factors tended to have positive outcomes in care despite stress and were characterised as 'resilient,' and children who lacked protective factors were more likely to develop emotional and behavioural problems under similar risk conditions. Psychologists suggested that for children in care resilience is a successful outcome of healthy adaptations during stressful life events (Rutter, 1987).

Furnivall (2011) argues that among other attributes which build resilience in children in care include a sense of humour, particularly the capacity to laugh at oneself, and the same applies to building resilient social work practitioners, so the social worker who uses humour is likely to be a more resilient social worker. Indeed I would suggest that having a sense of humour and being endearing (socially attractive) are the same attribute.

The Munro Report (2011) highlighted the importance of consistency in relationships and the need for workers to be emotionally resilient. Munro argued that amongst the factors which helped social workers develop resilience was the role of supervision, support and being able to express themselves emotionally (Munro Report, 2011). Hart et al. (2007) found that resilient people are people who have overcome adversity, and have long known the value of humour, therefore possessing a sense of humour is associated with resilience (Hart et al., 2007). Wolin and Wolin (1993) found that there was a connection between creativity and humour in people who are resilient, as humour counteracted the role of victim, and that laughing about your situation was creative as it produced new ways of being and enabled active resistance to adversity.

> *Social work trainees who are more adept at perceiving, appraising and expressing emotion, who are able to understand, analyse and utilise emotional knowledge, and who are able to regulate their emotions effectively appear to be more resilient to stress.*
>
> (Kinman and Grant, 2011, p. 262)

Wolin and Wolin (1993) found that in-group humour can help rid a group of negative emotions and strengthen social support among people who have come through trauma and challenges. Sharing one's survival of a difficult experience can mentor and encourage others. Psychiatrist Victor Frankl wrote in 1946 about surviving Nazi concentration camps that humour was a weapon for self-preservation, as it created the ability in humans to rise above the most awful of situations, if only for a few seconds (Frankl, 1946). Richards (2007) pointed out that in extremes of poverty and deprivation amongst the Nigerian people there are many characteristics in common: warmth, resilience and, above all, humour. Humour therefore provides a vital social function in relation to resilience to traumatic events. Moran and Hughes' (2006) study focused on whether students entering the social work profession already used humour or developed this as a mechanism for coping with stress, and they found that students who used humour to cope were less lonely and less depressed.

Humour and organisations

Organisational and workplace studies of behaviour have often been the source of many researchers' interests. There have also been considerable changes to the workplace, not least in terms of increased job insecurity, declining union membership and industrial action. As a result, the workplace has been the focus of many studies, some of which have focussed specifically on the use of humour in the workplace. Often located in ethnographic methodologies, such studies frequently disagreed on the overall effects and purposes of workplace humour.

Early studies of management e.g. Malone (1980) argued that humour was a prized, but underdeveloped resource at work, which managers could use to increase employees' satisfaction. Historically humour and laughter in the workplace was often seen as diametrically opposed to the efficient organisation and effective management of workers and in 'traditional management discourse, humour in the workplace was principally viewed as undermining productivity and subverting the maintenance of authority' (Karlsen and Villadsen, 2015, p. 517). In such an historical context humour was frowned upon as it had no place at work and should be constrained to leisure time.

However, driven in part by the reality of work behaviour and practices a growing body of research focussed on the use of humour in organisations, often supplemented with spurious claims that increased happiness equates to increased productivity. Some management 'gurus' made claims that happiness equates to job satisfaction which equates to company success (Sabrina, 2017). Whilst it is possible to be critical of such claims it is also true that this has been built on increased interest in humour and happiness at work. One of the first examples of workplace humour study was from Collinson (1988) which found that joking behaviour, even sexist, racist and other oppressive forms of humour, was the way in which work groups created their own identity in opposition to management.

Some research on humour in organisations has found that humour can be a double-edged sword (Cooper, 2008). Cooper (2008) pointed out that how humour is used in the workplace can be dependent on several factors, the perceptions of those involved and the subject of the humour. If the quality of the relationship with a manager is positive then the subordinate will interpret the use of humour positively and if the relationship is poor then subordinates are likely to attribute negative characteristics to a manager's use of humour (Cooper, 2008). The second factor is appropriateness of the humour, and whether the person using humour is 'in tune' with the culture of the organisation. Used appropriately humour can decrease conflict, release tension, increase morale or communicate a message (Cooper, 2008).

This was also linked in some studies to the relationship between the manager and their staff and levels of trust in the organisation (Wijewardena et al., 2017) and this is moderated through what Wijewardena et al. (2017) term as 'leader-member exchange relationships' between the employee and manager where such positive relationships are characterised by shared feelings of liking, loyalty, trust and emotional support.

Research into the use of humour in organisations reveals humour can be used in many ways. The use of humour at work is often seen as an artefact of the social system. When humour is carried out by those in power to those without power in the form of 'teasing' it can be done for 'getting the job done' and humour can be used by managers to reinforce power differentials (Cooper, 2008). Cooper (2008) found a correlation between managers' use of humour and the way they are perceived by their staff and a shared humorous experience allowed an individual to feel validated and drew workers and managers to feel closer (Cooper, 2008).

Cooper (2008) drew a distinction between vertical humour (humour between managers and subordinates) and horizontal humour (humour between co-workers). She also found that there were differences in how humour was used and its relationship to social attractiveness, and with those with high humour orientation being judged as the most socially attractive. 'Clowning humour' was judged more favourably than sarcasm, although this was mitigated by gender (men more likely to find sarcasm humorous than women) (Cooper, 2008). In developing a joking culture, workers created their own identity as distinct from management (Cooper, 2008). Cooper (2008) also suggested that humour can be a vehicle for *self-disclosure*, which is associated with liking and attraction within relationships. She argued that humour can be used to more safely disclose personal information, and that the humour one uses reveals something about oneself.

It is likely that when humour is used in the organisation by social workers two or more of the explanations for the humour may coalesce e.g. conveying superiority, managing incongruous feelings or relieving anxiety/stress (Gilgun and Sharma, 2011). There is potentially cyclical pattern to humour, i.e. when humour builds or detracts from a relationship, it encourages or discourages others from expressing humour. For example if someone feels themselves to be belittled at work, they are unlikely to respond positively. Humour is a form of social communication, which acts as a reinforcing or punishing event, as it *manipulates affect* (Cooper, 2008, p. 1101) and Holmes (2000) found that humour was an effective strategy for reducing offence.

Given that social work is constantly dealing with situations which can raise difficult emotions, humour can be a process which helps to manage these issues. Evidence from ethnographic research e.g. Locke (1996) supports this. Locke (1996) observed paediatricians' comedic acts with patients and their families, and found these interactions caused the family to like the doctor more. Could this also be true of social workers' encounters with children, where there is a need for children to feel liked and to like their social workers?

A key strategy in such group interaction, especially in dealing with possible tensions and dissent within the group, was the use of laughter, by both participants and moderators alike. Laughter was often used as an alternative to overt disagreement and the confrontation of difference. It forged bonds of support in the group while diffusing tensions around the topic under discussion and between those who disagreed.

(Brannen and Pattman, 2005, p. 526)

Robert and Wilbanks' (2012) *Wheel Model of Humour* emphasises the distinctively social nature of humour and indicates that 'humour-induced positive affect' results in transmission of emotion in groups, which in turn creates a climate that supports humour use and subsequent humour events. Robert and Wilbanks (2012) argue that humour events must be viewed as part of a cyclical and cumulative process whereby individual events have an incremental influence on affect, but also lay the foundation for additional humour events (Robert and Wilbanks, 2012). A key theoretical component of Robert and Wilbanks' (2012) theory is *emotional contagion*. Emotional contagion is a process whereby people imitate others' demonstrations of emotions such as facial expressions, speech, smiling and laughter which results in the actual experience of similar emotions (Robert and Wilbanks, 2012). As Carpenter (2011) suggested the 'climate' which social workers operate in is important. She suggests that a negative mindset and inability to feel positive about change which is so crucial to surviving in social work. In this sense the negative emotional contagion can have a harmful effect on the operation of social work teams, and the opposite can have a positive effect on teams.

Robert and Wilbanks (2012) suggest that although there are times when it might be possible to over-emphasise humour's potential positive effects, humour 'might be an unsung hero in peoples' day-to-day affective lives' (Robert and Wilbanks, 2012, p. 1093). But they also posit an important limitation that if humour can help stimulate and develop positive affect, how can it be that nobody has yet demonstrated this? (Robert and Wilbanks, 2012).

Like Cooper (2008) who pointed out that humour is used frequently in the workplace, humour and joke use has significant implications for interpersonal dynamics and relationships in organisations. Managers who utilise humour are perceived by employees as being more relationship oriented (Decker and Rotondo, 2001 cited in Cooper, 2008). Cooper argues that humour plays a significant role in the workplace relationships: 'humour dynamics can facilitate or detract from the formation of new relationships, as well as strengthen or destroy existing relationships' (Cooper, 2008, p. 1088). Cooper (2008) pointed out humour can be shared between employees or it can be targeted on an individual (including employers or service users). This is intentional use of humour and can include sarcasm, visual images, orchestrated jokes, storytelling or puns. It is also intended to cause a reaction.

Wijewardena et al. (2017) felt their study had practical implications for managers: 'humour is a work event that managers must manage and intentionally use to manage their employees' emotions for better functioning' (Wijewardena et al., 2017, p. 1332).

Stress and humour

Social workers are working with people under stress, often caused by issues such as poverty, child abuse, homelessness and discrimination, as a result it is not uncommon for social workers to experience stress (Moran and Hughes, 2006). Some writers have written about vicarious trauma and the ways in which social workers are themselves traumatised by working with adults and children who have experienced profound and enduring abuse and trauma (Conrad, 2011; VanDeusen and Way, 2006).

It is uncontentious to say that social work is a stressful job (Obholzer, 1997) and sharing jokes, using humour can help social workers cope with the stress of the work. Humour can be both a help in moderating the effects of stress and utilising humour can help others to deal with situations of extreme stress (Moran and Hughes, 2006). Sullivan (2000) argued that not only does 'gallows humour' serve the function of self-protection, it manages uncomfortable or derogatory thoughts about service users. As Mik-Meyer (2007) found social workers sometimes felt powerless to change their service users' lives. In consequence social workers used humour to deal with the frustrations and anxieties of the job, but this also became the first stage in moving on to constructive problem solving. Gilgun and Sharma (2011) found that case managers used humour to cope with and relieve stress. They found that stress included frustration and anger when parents were unwilling or incapable of handling their own issues and provide adequate care for their children. Gilgun and Sharma (2011) found that team meetings became the conduit for channelling frustration and the use of humour in this context went beyond stress relief and became a way of regulating negative emotions.

Moran and Hughes (2006) found that a person's sense of humour can be a mitigating factor in the effects of stress, and laughter can provide 'a form of control in uncontrollable situations by being self-affirming' (Moran and Hughes, 2006, p. 504). This is supported by studies including Freud's work (1960) and Lemma (2000) where a 'gallows humour' can help individuals wrest back some control in what might appear to be hopeless situations.

Could the use of humour aid social workers to be more confident in their practice? Barron (1999) points out that a reduced capacity for humour 'impoverishes our psychic life.' Stuber et al. (2007) found that humour can help children and adolescents tolerate painful medical procedures. Other research that laughter has long been viewed as 'good medicine' for a variety of illnesses (Bennett and Lengacher, 2006). It has long been part of folk mythology that being happy makes you feel good. The relationship between humour, physical and mental health in Chapter 5.

The idea here is that using humour in the workplace helps the worker not only to survive, but to thrive. Satyamurti (1981) found that social workers used a variety of strategies to make their work more tolerable, and she found these strategies were both individual and collective, practical and conceptual. Social workers shaped their situations in the context of their main relationships.

Satyamurti (1981) also suggests that social workers experienced meaningless in their work, which contradicts with the public perception of social workers as a vocation, with a capacity of creative and helpful human relationships. She also noted that a process of deskilling, i.e. a move away from traditional casework was occurring in 1981, and social workers rarely experienced what she termed as successful work. Nearly 30 years later Overell (2008) reinforced this and suggests that there is a dilemma in the way modern work is constructed, as work is an arena for self-realisation, but is also increasingly alienating and oppressive, as we come to work to do something good, but find that we do not and find little kinship or solidarity. Uttarkar (2008) found in that staff in a mental health team she studied utilised humour to mask fear and to evade discussing its impact on staff. The team depended on humorous interludes to seek relief from relentless discussions about the painful experiences of their patients (Uttarkar, 2008).

Moran and Hughes (2006) found that social work students did not see humour as one of their coping strategies in contrast to experienced social workers who relied on humour to manage or cope with stress. Moran and Hughes (2006) expected to find that a positive attitude to humour would be associated with other positive measures for managing stress, but found that attitudes to humour amongst social work students revealed little relationship to stress or health levels. They also found that some humour requires 'permission, which may be explicitly or implicitly sought' (Moran and Hughes, 2006, p. 512) as social work students felt guarded about when it was appropriate to talk about humour regarding social work. Moran and Hughes (2006) cited the example of a mental health team where the team shared laughter over the 'funny side of service users behaviours,' although the practitioner needed to reassure the student this did not carry over to face to face interactions. It is this contradictory nature of humour use that has been found in other research e.g. participants in Sullivan's (2000) study found that by talking about humour and its uses they were discovering something new about their interactions with service users. Lemma (2000) claimed that a judicious use of humour can aid the development of a positive atmosphere during therapy, and it is possible to suggest that this could also be true of social work, particularly in supervision. It can be uncomfortable for a social worker to become aware they have responded in a prejudiced manner, but unwanted prejudicial thoughts have a better chance of being managed if they are expressed (Sullivan, 2000).

Chiller and Crisp (2012) point out that social work as well as being emotionally demanding and stressful, is lacking in rewards. Under-resourcing, high staff turnover, high caseloads often translates into poor morale and negative organisational culture (Chiller and Crisp, 2012). As a result, the average working life for a social worker is only eight years compared to 15 for nurses (Chiller and Crisp, 2012). Chiller and Crisp (2012) argue that humour can be (alongside resilience, mindfulness and emotional intelligence) a strategy for improving the longevity of social workers and their own self-care. I have found that a study of the uses of humour in relation to social work could therefore provide insight into the employment retention of social workers.

Jokes and the use of humour by social workers and their colleagues are cited as one of their most common coping mechanisms (Moran and Hughes, 2006). Gilgun and Sharma (2011) stated that social workers rarely confronted humourous situations in their everyday practice but suggest that humour may help 'social workers deal more effectively with difficult situations' (Gilgun and Sharma, 2011, p. 2).

As the relief theory of humour suggested, jokes and humour can help people tolerate and face adversity, and as anthropological studies indicate in-group humour can help acknowledge and dispel negative emotions. Humour can be a strengthening factor in social support among people who have come through trauma and challenges, and it can also be a way for people who have survived a difficult experience to support and encourage those who are still going through it.

Social work identity

Brannen and Pattman (2005) carried out a study in social services departments and found that laughter can stand in for criticism in social work and that laughter served to create solidarity between social workers. Brannen and Pattman (2005) also found that laughter and humour are important ways of signalling tension and dissension in social work teams, but also of managing it and rendering a situation less threatening.

> *In another group (FG9), one of the participants was supervised but not managed by another participant. Here, humour was used to manage the uncertainty that arose from the ambiguous situation caused by the presence of a more senior colleague. The participant joked that her supervisor was 'very nice,' and responding to this, the supervisor quipped she 'had' to say this. When criticisms were voiced in this group and levelled against various managers and supervisors (though not the one participating) the supervisor was critical of her own manager though less vocal than others in the group. Her quietness reflected her discomfort about being in a supervisory position. But when she did speak she seemed to be trying to come across as being 'on the side' of the staff.*
>
> (Brannen and Pattman, 2005, p. 529)

In some groups Brannen and Pattman (2005) looked at participants became silent accomplices and 'their criticism of (absent) managers which was only hinted at through laughter among participants' (Brannen and Pattman, 2005, p. 535). So laughter and humour can be used to single dissent and criticism in ways considered safe by practitioners.

Humour can also be a mechanism through which groups and individuals create their own identity. Professional identity can refer to the characteristics, intentions and principles through which an individual defines him or herself (McSweeney, 2012), whilst for others professional identity is developed in the community of practice to which it belongs (Wenger, 1998). Lockyer and Pickering (2009) argue that identity is linked to the use of humour. Stereotypes exist in relation to social work and as noted in Chapter 4 *Clare in the Community* and *Damned* utilise stereotypes for comedic effect. Other authors have also suggested that humour is central to social workers identity, e.g. Drinkwater (2011) wrote that among the personal qualities that make a good social worker, he would add a sense of humour, alongside empathy, integrity, objectivity and perseverance. He suggested that humour can be useful with a joyless or 'curmudgeonly' colleague or to 'break down barriers with steely clients' (Drinkwater, 2011). Rogowski (2011) too suggested that frontline social workers, particularly child protection social workers, have always used humour, often gallows humour, as a way of managing their day-to-day work. However, he also acknowledged that there may be times where humour is used inappropriately, and when social workers are disrespectful of service users, but he excused this by commenting that social workers are 'not robots working in a sterile environment' (Rogowski, 2011).

The social workers' 'tools of their trade' are their personality, temperament, intelligence and concern for the person they are working with (Griffiths, 2017). Sullivan (2000) argued that humour used by social workers emphasised the difference between them and service users in terms of status, saneness, intelligence and knowledge. In this sense social workers saw themselves as superior to their clients. Sullivan's (2000) findings are important as they suggest that social workers could be practicing in unethical and oppressive ways. Importantly the HCPC Standards of Proficiency require social workers to *be able to understand the emotional dynamics of interactions with service users and carers* (HCPC, 2017 9.10), so it's possible that oppressive workplace use of humour could be a concern for the profession.

One could posit that humour is a critical part of emotional dynamics, which impact on our interactions with service users, and provides us with insight into modern social work practice. Social work and humour has been studied before and it has been found that humour and the sharing of jokes and funny stories can build resilience in social work teams (Siporin, 1984; Witkin, 1999; Moran and Hughes, 2006; and Gilgun and Sharma, 2011). Gilgun and Sharma's (2011) study found positive aspects of humour use, as the social workers in their study used humour to regulate anxiety, exasperation and fear. In addition, the social workers often used humour to regulate their emotions, to problem solve and to express liking of service users.

Practice issues

This section considers the implications for practice from the arguments contained in this chapter, e.g. what does an analysis of humour in relation to the workplace and organisations studies suggest for social work practice? As Moran and Hughes (2006) concluded using 'humour socially may have some stress-moderating properties' Moran and Hughes, 2006, p. 512).

In January 2018 an article appeared in the Community Care that reported an HCPC panel a social worker had been struck off for putting children at risk had her spirit 'broken' by service restructure. '[The supervisor] told the panel that the registrant's personality changed from someone who had a sense of humour and was cheerful, whose work was good and who was focussed, keen and who paid attention to detail, to someone who was distracted, whose spirit was broken, and who was not as sharp as she was before' (Stevenson, 2018). It significant here for practice issues that a sense of humour is the mechanism though which she was measured in terms of her ability to continue to perform effectively in her role. Whilst there was criticism of the council and their role, it is significant that loss of a sense of humour was a factor in this social worker eventually being struck off. Whilst this may seem unfair and a further tool to criticise social workers with, it also illustrates the importance of humour or at least a sense of humour in the workplace in terms of managing and maintaining safe working practices for social workers and people receiving services.

If used selectively, sensitively and appropriately, humour can be seen alongside other mechanisms as a positive method of dealing with stress (Collins et al., 2008; Drinkwater, 2011; Rogowski, 2011). There is evidence of the lack of humour and levels of unhappiness in the social work profession and some authors suggest this begins in social work training (Tobin and Carson, 1994; Collins et al., 2008).

Humour plays a role in the work environment and can be linked to job satisfaction and in-group solidarity as it helps social workers bond together through shared experiences and another commentator on line stated:

> *Humour is essential and lucky for me plentiful could not get through the day without it.*

(Jordan, 2015)

Social workers interviewed as part of my research into humour use (Jordan, 2015) indicated that humour was central to social work practice and important for creating and maintaining work relationships. In the following extract a practitioner made explicit reference to how humour can enrich social workers' working life and make the task of social work more tolerable for social workers themselves:

> *I've enjoyed working in teams . . . where I have had job satisfaction, or where I've been happy there and there's been banter . . . so I think [humour] . . . kind of oils the wheels of human interaction, I think whatever kind of work situation you are involved with I think humour kind of helps jolly the day along really I guess, for one thing . . . . want of a better thing really . . . and . . . . in that sense I see it as a kind of cement or a glue in relationships, without which life would be all the more impoverished. Relationships arguably more likely to become unstuck!*

(Jordan, 2015)

The refrain that humour helped social workers make relationships was a theme which arose in on-line comments:

> *Our team is full of banter and jokes and it's partly why I love the job. Yes, the humour is sometimes dark and would not be particularly amusing to those not working in the field but it helps you get through the stress and helps you bond as a team. We do often get told off though the laughing too much.*

(Jordan, 2015)

Conclusion

It is likely that particular forms of organisation management create distinct forms of humour or joking behaviour. Thompson and Ackroyd (1995) found that satirical humour is common in the contemporary workplace, where there is often deep distrust of management motives, so it is not perhaps surprising that this form of humour should also be found in social work workplaces.

Humour can be used in the workplace to bully and oppress and as shown in the example from practice when a worker 'loses their sense of humour' because of unwanted organisational changes and humour can be a measure of the health not only of an individual but the social work team. If this however was to become another management or supervisory tool used insensitively e.g. 'I'm worried

about you because you didn't laugh at my joke' then this would also be unlikely
to improve a team's functioning. However used sensitively and by staff or managers it can also be a sign of a resilient worker.

A potentially fruitful area of research is in the relationship between humour,
staff turnover and absenteeism. Robert and Wilbanks (2012) suggest that humour
might play a role in decreasing these phenomena and increasing the incidence
of positive humour events might reduce the probability of 'impulsive quitting'
owing to negative affective states and by aiding in the development of strong
friendships at work, which have been associated with lower turnover rates (Robert
and Wilbanks, 2012).

References

Barron, J. W. (Ed.) (1999) *Humour and Psyche: Psychoanalytic Perspectives* Hillsdale, NJ:
The Analytic Press.

Bennett, M. P. and Lengacher, C. A. (2006) Humour and laughter may influence health II:
Complementary therapies and humour in a clinical population *Evidence-based Complementary and Alternative Medicine* 3, 187–190.

Bennett, O. (2013) Beaten toddler Keanu Williams 'invisible' to authorities before he died
Daily Express www.express.co.uk/news/uk/434084/Beaten-toddler-Keanu-Williams-invisible-to-authorities-before-he-died (accessed 6/10/13).

Billig, M. (2005a) *Laughter and Ridicule towards a Social Critique of Humour* London:
Sage Publications.

Brannen, J. and Pattman, R. (2005) Work-family matters in the workplace: The use of focus
groups in a study of a UK social services department *Qualitative Research* 5, 523. Sage
Publications.

Carden, I. (2003) On humour and pathology: The role of paradox and absurdity for ideological survival *Anthropology & Medicine* 10(1), 115–142.

Carpenter, J. (2011) *A Response to Drinkwater, M. (2011) On Reflection: Mark Drinkwater
on Humour in Social Work* https://mypotentio.bloomfire.com/posts/6593-humour-in-social-work/public

Chiller, P. and Crisp, B. R. (2012) Sticking around: Why and how some social workers stay
in the profession *Practice: Social Work in Action* 24(4), 211–224.

Collinson, D. L. (1988) Engineering humour: Masculinity, joking and conflict in shop floor
relations *Organization Studies* 9(2), 181–199.

Collins, S (2008) *Statutory Social Workers: Stress, Job Satisfaction, Coping, Social Support and Individual Differences* British Journal of Social Work (2008) 38,
1173–1193.

Conrad, D. (2011) Secondary trauma and caring professionals: Understanding it's impact
and taking steps to protect yourself *The Link* 20(2), 1–5.

Cooper, C. (2008) Elucidating the bonds of workplace humor: A relational process model
Human Relations 61, 1087–1115.

Drinkwater, M. (2011) *On Reflection: Mark Drinkwater on humour in social work* http://
www.communitycare.co.uk/2011/01/06/why-humour-is-so-important-in-the-social-work-workplace/(accessed 06/01/2011).

Elkes, N. (2013) *Birmingham child protection agencies failed two-year-old Keanu
Williams – report Birmingham Mail* http://www.birminghammail.co.uk/news/local-news/birmingham-child-protection-failed-keanu-6131773 (accessed 6/10/13).

Ferguson, I. (2008) *Reclaiming Social Work: Challenging Neo-Liberalism and Promoting Social Justice* London: Sage Publications.

Frankl, V. (1946) *Man's Search for Meaning* London: Hodder and Stoughton.

Freud, S. (1960) *Jokes and Their Relation to the Unconscious* London: Routledge & Kegan Paul.

Furnivall, J. (2011) *Guide to developing and maintaining resilience in residential child care* www.ccinform.co.uk/articles/2011/11/07/6507/guide+to+developing+and+maintaining+resilience+in+residential+child.html?

Gilgun, J. F. and Sharma, A. (2011) The uses of humour in case management with high-risk children and their families *British Journal of Social Work* 1–18.

Griffiths, M. (2017) *The Challenge of Existential Social Work Practice* London: Palgrave Macmillan.

Hare, I. (2004) Defining social work for the 21st century: The international federation of social workers' revised definition of social work *International Social Work* 47(3), 407–424.

Harlow, E. (2003) New managerialism, social services departments and social work practice today *Practice* 15(2), 29–44.

Harlow, E., Berg, E., Barry, J. and Chandler, J. (2013) Neoliberalism, managerialism and reconfiguring of social work in Sweden and the United Kingdom *Organisation* 20(4), 534–550.

Harris, J. (2008) State social work: Constructing the present from moments in the past *British Journal of Social Work* 38(4), 662–679.

Hart, A. and Blincow, D. with Thomas, H. (2007) *Resilient Therapy: Working with children and families* London: Routledge

HCPC (2017) *Standards of proficiency: Social workers in England London* Published by Health Care Professions Council and https://www.hcpc-uk.org/

Holmes, J. (2000) Politeness, power and provocation: How humour functions in the workplace *Discourse Studies* 2, 159–185.

Hope-Hailey, V., Dietz, G. and Searle, R. (2012) *Where Has All the Trust Gone?* London: Chartered Institute of Personnel and Development (CPID).

Horner, N. (2007) *What Is Social Work* Exeter: Learning Matters.

Jordan, B. (1984) *Invitation to Social Work* London: Martin Robertson & Co Ltd.

Jordan, S. (2015) *That joke isn't funny anymore: Humour, jokes and their relationship to social work* London: University of East London. Professional doctorate thesis.

Karlsen, M. P. and Villadsen, K. (2015) Laughing for real? Humour, management power and subversion *Ephemera* 15(3), 513–535. ISSN 1473-2866 (Online) ISSN 2052-1499 (Print) www.ephemerajournal.org

Kinman, G. and Grant, L. (2011) Exploring stress resilience in trainee social workers: The role of emotional and social competencies *British Journal of Social Work* 41, 261–275. DOI: 10.1093/bjsw/bcq088 (accessed 24/8/10).

Legood, A., McGrath, M., Searle, R. and Lee, A. (2016) Exploring how social workers experience and cope with public perception of their profession *British Journal of Social Work* 46, 1872–1889 (accessed 13/1/16).

Lemma, A. (2000) *Humour on the Couch* London: Whurr.

Lewis, B. (2009) *Hammer and Tickle* London: Orion Books.

Locke, K. (1996) A funny thing happened! The management of consumer emotions in service encounters *Organization Science* 7(1), 40–49.

Lockyer, S. and Pickering, M. (Eds.) (2009) *Beyond a Joke: The Limits of Humour* Basingstoke: Palgrave Macmillan.

Lymbery, M. (2005) *Social Work with Older People* London: Sage Publications.

Malone, P. B. (1980) Humor: A double-edged tool for today's managers *Academy of Management Review* 5(3), 357–360.

Masten, A. S. and Garmezy, N. (1985) Risk, vulnerability and protective factors in developmental psychopathology in Lahey, B. B. and Kazdin, A. E. (Eds.), *Advances in Clinical Child Psychology* (Vol. 8, pp. 1–512) New York: Plenum.

McSweeney, F. (2012) Student, practitioner, or both? Separation and integration of identities in professional social care education *Social Work Education* 31(3), 364–382.

Mik-Meyer, N. (2007) Interpersonal relations or jokes of social structure? Laughter in social work *Qualitative Social Work* 6(9).

Moran, C. C. and Hughes, L. P. (2006) Coping with stress: Social work students and humour *Social Work Education* 25(5), August, 501–517.

Mullaly, B. (1997) *Structural Social Work: Ideology, Theory and Practice* (2nd ed.) Toronto, ON: Oxford University Press.

Munro, E. (2011) *The Munro Review of Child Protection: Final Report a Child-Centred system* London: HMSO.

Obholzer, A. (1997) *The Unconscious at Work: Individual and Organizational Stress in the Human Services*, eds A. Obholzer & V. Zagier Roberts, London, Routledge.

Overell, S. (2008) *Inwardness: The Rise of Meaningful Work Provocation* London: The Work Foundation.

Payne, M. (1996) *What Is Professional Social Work?* Birmingham: Venture Press.

Richards, K. (2007) Warmth, resilience and humour *Financial Times* July 12.

Robert, C. and Wilbanks, J. E. (2012) The wheel model of humor: Humour events and affect in organizations *Human Relations* 65, 1071 originally published online 27 April 2012. DOI: 10.1177/0018726711433133Rogowski, S. (2011) You've got to laugh: Frontline focus *Community Care* 3.

Rogowski, S. (2012) Social work with children and families: Challenges and possibilities in the neo-liberal world *British Journal of Social Work* 42(5), 921–940.Rutter, M. (1987) Psychosocial resilience and protective mechanisms *American Journal of Orthopsychiatry* 57, 316–331.

Sabrina, D. (2017) Happy employees are productive employees *Huffington Post* March 4 03:17 pm ET www.huffingtonpost.com/entry/happy-employees-are-productive-employees_us_5906380fe4b084f59b49fa46

Satyamurti, C. (1981) *Occupational Survival* Oxford: Blackwell.

Scheper-Hughes, N. (1993) *Death without Weeping: The Violence of Everyday Life in Brazil* Berkeley: University of California Press.

Seed, P. (1973) *Expansion of Social Work in Britain* London: Routledge & Kegan Paul.

Siporin, M. (1984) '*Have you heard the one about social work humor?*', Social Casework: The Journal of Contemporary Social Work, 65, pp. 459–64.

Stevenson, L. (2018) Social worker struck off for putting children at risk had spirit 'broken' by service restructure, HCPC hears A HCPC panel said actions by the social worker indicated poor performance 'was intentional' *Community Care* January 17.

Stuber, M., Dunay Hilber, S., Libman Mintzer, L., Castaneda, M., Glover, D. and Zeltzer, L. (2007) Laughter, humour and pain perception in children: A pilot study *Evidence-based Complementary and Alternative Medicine* 6(2), 271–276.

Sullivan, E. (2000) Gallows humour in social work practice: An issue for supervision and reflexivity *Practice* 12(2), 45–54.

Thompson, N. (2005) *Understanding Social Work: Preparing for Practice* (2nd ed.) Basingstoke: Palgrave Macmillan.

Thompson, P. and Ackroyd, S. (1995) All quiet on the workplace front? A critique of recent trends in British industrial sociology *Sociology* 29(4), 610–633.

Tobin, J. and Carson, J. (1994) *Stress and the student social worker* Social Work and Social Sciences Review, 5(3), pp. 246–56.

Uttarkar, V. (2008) *An investigation into staff experiences of working in the community with hard to reach severely mentally ill people*. London, University of East London. Thesis submitted for the award of Professional Doctorate in Social Work, at the University of East London in Collaboration with the Tavistock Clinic October.

VanDeusen, K. M. and Way, I. (2006) Vicarious trauma: An exploratory study of the impact of providing sexual abuse treatment on clinicians' trust and intimacy *Journal of Child Sexual Abuse* 15(1), 69–86.

Warner, J. (2017) Social work must challenge political and public anxieties about child protection *Community Care* www.communitycare.co.uk/2017/11/15/social-work-must-challenge-political-public-anxieties-child-protection/ (accessed 15/11/17).

Wenger, A. (1998) *Communities of Practice: Learning, Meaning, and Identity* Cambridge: Cambridge University Press.

Wijewardena, N., Härtel, C. E. J. and Samaratunge, R. (2017) Using humor and boosting emotions: An affect-based study of managerial humor, employees' emotions and psychological capital *Human Relations* 70(11), 1316–1341. DOI: 10.1177/0018726717691809j

Witkin, S. L. (1999) '*Taking humour seriously*' Social Work, 44(2), pp. 101–4.

Wolin, S. J. and Wolin, S. (1993) *The Resilient Self* New York: Villard Books.

4 The media

'Equal opportunity approaches to offensiveness'

Q: What's the difference between a real social worker and a bogus social worker?
A: You receive a visit from a bogus social worker

Introduction

In Britain we are a culture which has placed great value on various forms of humour including satire and pride ourselves on our shared sense of humour, for example in writing about the depressing spectacle of the government's machinations over Brexit: 'the knowledge that the British enthusiasm for satire – it takes a certain type to carry on laughing as you're being driven over the cliff edge – shows no sign of diminishing' (Crace, 2017). The people who make us laugh are highly paid and their popularity fills stadiums (Dessau, 2012), in this respect we love to laugh and have fun, even in the bleakest of times, and cartoons, jokes and comedies can be the source for this fun. I must admit to a fondness for cartoons, which I feel can be a coping mechanism for many to lighten the dullness of existence, and I think occasionally can communicate deeper emotional truths than other forms of fiction, in a few panels or limited dialogue.

Modern popular media such as the newspapers, television, radio, satellite and pay for view channels like Netflix are a key part of how most people access entertainment. Whilst increasing numbers of people are starting to watch and listen to pay for view channels like Netflix, the mainstream TV stations such as BBC, ITV, Channel 4 and 5 remain the most popular ways of accessing the media. Popular dramas such as EastEnders, Coronation Street or Holby City, infrequently feature social workers in episodes as superficial characters playing minor roles. This chapter considers the humourous representations of social workers on TV and in cartoons, both in newspapers and the professional journal, Community Care.

Media representations and the public image of social work

Every professional group, in whatever walk of life, has a public image: Police officers are traditionally portrayed as being tall, with flat feet. Social workers, on the other hand, have beards and tweed jackets. As well as the physical appearances of our respective professions, the public also attribute certain behavioural characteristics: Social workers are supportive but not very useful. Police officers are domineering and oppressive. We are all aware of these caricatures or exaggerated stereotypes. It benefits us to examine the basis of these views and it is true that perceptions are based on limited information and knowledge. It is a fact of policing and social work that, by definition, we encounter members of the public only when things have gone wrong in their lives. For the rest of the population, who are fortunate enough in that they have had little or no need for our services, much of their information, certainly with regard to the police, derives from the media.

(Strang, 2003, p. 1)

The media are a primary source of information regarding social services too (Legood et al., 2016). One example of regular social work humour in the media is *Clare in the Community*. *Clare* was originally created by Harry Venning in 1994 for the Guardian newspaper as a strip cartoon about the exploits of a social worker and she continues to appear in the paper on a weekly basis. In 2004 the cartoon was adapted for BBC Radio 4 half an hour comedy series and is now in its eleventh series. The humour in *Clare* centres on *Clare* as a social worker who meddles in other people's problems rather than deal with her own. Other examples of social work humour include 'Fran' who is a cartoonist who draws cartoons regularly published in the Community Care magazine, and another example of social work humour in this section is 'Damned,' a comedy created by Jo Brand and shown on Channel 4. The series was set in the office of a children's social work team.

Social work is concerned with its image in the popular media and Tower (2000) found that in America 'defamatory images of the profession are commonplace, especially in the popular media' (Tower, 2000, p. 575). In the UK too, there has also been concern about how social workers are shown in the media, for example Ruth Allen (Chair of BASW) was concerned about the portrayal of social workers in an EastEnders episode and described the depiction as a 'cardboard cut-out' portrayal of a social worker (Stevenson, 2017). Eborall and Garmeson (2001) found that in the 1990s the reporting of social work care issues has been almost unrelentingly critical. Miller and Bartlett (2004) found that in terms of influencing the public perceptions of social work, the media is considered to be the single most important factor, not least by social workers.

Tower (2000) argued that in order to counter negative depictions of social workers 'social workers should be using television if they hope to increase knowledge and change attitudes about the profession and its constituents' (Tower, 2000, p. 576). For Legood et al. (2016) the impact of damaging representations of social workers has adverse effects on recruitment and the morale of the profession and found that for the majority of participants in their study

simply experiencing a 'good news story' about their profession would be extremely welcome and uncharacteristic of their present portrayal. Some efforts made in recent years were acknowledged, such as those that shine a more positive light on the role of social workers, through various more 'realistic' documentaries and the provision of more positive examples on television. However, others acknowledged that changing public perception via the media would not be an easy task: ' . . . with the British media you do feel that actually the newspapers don't necessarily want to be publishing positive stories . . . . so I think it is a bit of an uphill battle.'

(Legood et al., 2016, p. 1884)

This is reinforced by Tower in her earlier study: 'Television is a vehicle for advocacy, therapy, public information, peer support, recruitment, and qualitative or quantitative research. This medium is the most pervasive form of communication in the world, yet, social work's presence and influence is almost nonexistent' (Tower, 2000, p. 585). This need for social workers to have positive stories about them also extended to service users, who suggested that what enabled social workers to do a good job was (amongst other issues such as having enough time to do the work and having a good attitude) that social work itself should have 'good publicity' (Perry, 2005). Legood et al. (2016) contend that in terms of staff recruitment, retention and an 'individual social workers' sense of pride and identity' being perceived positively by the public as a profession is important for the professions overall effectiveness and vitality (Legood et al., 2016, p. 1873).

There is an intuitive appeal then to seeing social workers in this most pervasive form of communication, albeit for fictional comedic purposes. So whilst 'television drama and comedy has churned out a number of forgettable, flimsily drawn social workers' (Wain and Stevenson, 2016) in the context of undesirable and at times damaging representations of social workers, in real or fictional settings on television and elsewhere, there is a value in considering the humourous or comic depictions of social workers in the media, as it is another vehicle through which people can see the presence of social workers. So, this chapter considers cartoon and comedy series which feature social work, as social workers often appear in fiction as side characters in comedies such as *Harry Enfield*, in cartoons such as *the Simpsons*, or as central characters in *Clare in the Community* and *Damned*. Frequently such cartoons are crude stereotypes created for popular consumption, but because they are the ways in which people encounter social workers they are examined here.

Single cartoon portrayals

Whilst it can be said that social work has an image problem, at least in the media portrayal of social workers, social workers have been occasional figures in cartoons. One of the earliest representations I discovered was a single frame cartoon by Ken Pyne which was published in *Punch* magazine in November 1980. It's an example of gallows humour, which depicts two men perched on a balcony ledge about to jump with one man on another man's shoulders. From a window a police officer leans out and says to another man, 'apparently the one underneath is the other one's social worker.' Social workers feature infrequently in *Punch* and some of the cartoons utilised humour based on literature e.g. with Fagin from Oliver Twist being reassured to hear that 'social services have determined that the children in your care are not at risk' (January 1991) or a social worker informing Winnie the Pooh and Piglet 'Hush, hush, whisper who dare Christopher Robin's been put into care' (June 1991).

Social workers in TV series

Whilst the cartoon representations examined in this section are primarily American in origin, they are included in this section because they have featured on British television.

The Simpsons

Whilst this chapter focuses on the representations of social workers in humorous fictional forms, it's worth noting that social workers have featured in cartoons and humourous series aimed at children. *The Simpsons* was one of the most popular shows on television with a peak of 33.6 million American viewers in 1990 with the episode 'Bart Gets an F,' which was the highest-rated episode in *Simpsons* history (Fast Company, 2009). In the UK the show has regularly attracted over 3 million viewers (Byrne, 2004), although in recent years this has been declining.

Social workers notably featured in several Simpsons episodes including 'Home Sweet Homediddly-Dum-Doodily' in 1996 when Homer and Marge are found to have neglected the children, who are placed in the foster care with the Flanders family by social workers from the Child Welfare Board. This is a cautionary tale featuring 'trendy social workers' and the morally upright Flanders family who are contrasted with the imperfections of Homer and Marge (Delingpole, 2010). The episode demonstrates that, in the end, the children are better off remaining with their 'good-enough' parents despite their faults, as they clearly love them. The representation of the social workers in this episode is of efficient bureaucrats, who leave their office in a large van to Batman-like music as they speed to the Simpsons' home. In the home the social workers are stern and critical of the parenting the children receive and immediately remove the children to a van, without any court order or legal process.

In an episode shown in 2002 entitled 'Brawl in the Family,' the seventh episode of Season 13, a social worker named Gabriel is drafted in to help the family after they have tried to strangle each other. This episode includes the social worker saying to Homer 'You're a drunken, childish buffoon. Which is society's fault because it's your fault, Homer.' In the episode the social worker loses his temper and is disgusted when he meets 'Homer's Vegas wife' and gives up after trying to change the family's behaviour. This episode was criticised for being unrealistic and having a ridiculous set-up (Martin, 2010). The only other time a social worker appears is as an unnamed female social worker who works at the Springfield Orphanage and briefly features in the eighth episode of Season 21 in 'O Brother, Where Bart Thou?'

King of the Hill

King of the Hill was a cartoon series created by Mike Judge and Greg Daniels set in Texas featuring a character called Hank Hill, his wife Peggy and son Bobby. The series aired on television from 1997 until 2010 (Shattuck, 2009). Notably the pilot episode featured a social worker named Anthony Page who worked for the 'Arlen County Child Protective Services.' In this episode the social worker investigated a black eye to Bobby, caused by a baseball, but rumoured by neighbours to have been caused by child abuse from Hank. In the episode the social worker uses jargon 'Mr. Hill, I feel that you're coming from an anger mindset, and if you're projecting your anger onto me, it gives me grave concerns as to how you facilitate your son's growth in private' and incompetently investigates the allegations until he is removed from the case by his manager. The depiction of the social worker is negative, as his behaviour is condescending, invasive and inappropriate.

Family Guy

Family Guy is series created by Seth McFarlane. The irreverent humour in the show has been criticised on the grounds of crudeness, obscenity and including offensive jokes about such subjects as terrorist atrocities, mentally disabled people and sexual assault, but some critics found it had an 'equal opportunity . . . [approach to] offensiveness that it is difficult to actively disapprove of' (Power, 2015). The show has attracted over half a million viewers in the UK (Hooton, 2015).

In an early season episode 'Love They Trophy' 'Sandy Belford' is a social worker from child services who removes Stewie from the family home and places him in foster care and is seen making such basic errors as interviewing the neighbours, who are biased in their comments, but neglecting to establish who was actually the mother of Stewie.

Two different male social workers appeared in one *Family Guy* episode, from 2005 entitled 'Petarded.' In the episode Peter gets a 'state appointed inspirational social worker' to help him. The enthusiastic and energetic Vern spends a lot of time giving Peter high fives and, at one point, is seen bathing him. According to Internet Movie Database Vern was based on the character Vern (played by Michael D. Roberts) in the film *Rain Man* (IMDb, 2018). In the same episode another social worker 'Agent Jessup' from *Child Services* informs Peter Griffin 'I'm here to take your kids away because you're mentally unfit to take care of them.' Agent Jessup looks more like an FBI agent than a social worker and in court comments: 'Peter Griffin you've inspired me . . . to distrust all mentally challenged parents.' In one single episode we see two contrasting caricatures of social workers, an enthusiastic and accepting social worker for learning disabled people, as opposed to judgmental and critical social worker from child protection.

South Park

Created by Matt Stone and Trey Parker *South Park* is a cartoon series centred on four nine-year-old boys which first aired on television in 1997 after Stone and Parker had created two animated short films a few years earlier and attracts high audience numbers (Bradley, 2016). *South Park* became infamous for its humour based on transgressing and dismantling serious and important cultural topics (Cogan, 2012). Significantly from a social work perspective, the show also contained characters with disabilities, *Timmy* and *Jimmy*. Chemers and Karamanos (2012) suggest that including characters with disabilities shows the power of inclusion to break down barriers, as 'they are neither 'special' in the sense of being put above traditional jokes nor special in the sense that they are singled out for ridicule' (Chemers and Karamanos, 2012, p. 57). Social workers appeared infrequently in *South Park* episodes and in the *South Park Bigger Longer & Uncut* movie in 1999.

A more significant example of a social worker, which was first aired in 2011, was entitled 'The Poor Kid.' In this episode social workers remove Kenny and his siblings from his parents and put them in foster care, after police discover a meth lab in the home. This episode raised questions for practitioners (Wright, 2011), as the children are interviewed by Mr. Adams, a case worker who wants to be a comedian and uses the session to practice his routine and poke fun at the Penn State sex abuse scandal, and it was this portrayal of a social workers who joked about abuse that raised criticism from some practitioners.

Fran Orford (Fran)

I first came across Fran Orford's cartoons in Community Care in 1988 and for many years he continued to produce cartoons about social work (and still does). Fran created in his words thousands of cartoon strips for not only Community Care, but a range of professional magazines, but also has a particular relationship with social work as he originally worked as a social worker and was a manager for a project for the National Childrens Home so has 'tremendous respect for the profession and most of the people who work in it, not just for their dedication, but for their ability to get anything done when strangled by bureaucracy, beaten up by a badly informed public and excoriated by politicians who should know better' (Orford, 2017). Fran is widely published, and he has cartoons published in over 70 publications in the UK and abroad including *Private Eye*, the *Times*, *New Statesman* as well as having had a regular strip in both the *Observer* and the *Daily Telegraph*.

In terms of the social work humour exemplified in Fran's cartoons, we frequently see two or three office workers in a single frame cartoon e.g. in one cartoon a man precariously holds a stack of file while another telling a female colleague 'It's our latest caseload management system, we just keep adding files until he falls over.' In another above a strapline 'Report suggest social workers spend 60% of time on admin,' a female colleague asks a male colleague 'do you remember when all we had to worry about was client care?' Fran's cartoons reflect a largely benign approach to humour. Fran developed his own idiosyncratic style for practical purposes: 'I'm not trained in anyway, in terms of illustration or art, so for me I developed a certain style and it's just easier to draw certain characters. . . . The same characters pop-up because they are just easier to draw' (Orford, 2018).

Fran's cartoons often focus on the idiosyncrasy of the social work workplace and he described an example which appeared in the Community Care magazine: 'the manager is saying: I've noticed you've been in the office all the time, and it's an issue I'm really concerned about, because you seem to be here 15 hours a day.' Then the punchline is: 'so I'm going to have to think about you charging you for the utilities' (Orford, 2018). The key aspect of the humour, for Fran, was to take some part of reality (management practices) and exaggerate it, so that there would be

> *social workers who would think 'that's not too insane, I can imagine them saying that to me'. . . you want [the joke] to be not mainstream, to have something quirky, something different about it, that's absolutely the case but if it's completely out of the park, then it doesn't work. If people can't get an entrance in, 'that's the kind of bloody thing they would do.' This guy coming in and saying 'we want you to work in a paperless environment, and this is how we do it, and then hands you five folders full of paperwork' – it happens.*

> (Orford, 2018)

Fran's work depicts what might be called a type of social work humour and indicated that as he had worked with a range of professional groups, different professions had distinct approaches to how they perceived humour made at them:

> *I did the law magazine for a couple of decades and you could absolutely rip the soul out of lawyers, they loved it, you could really take the mickey – you could present them as avaricious, selfish, ego driven eejiots and they liked it. I also worked for a magazine which was aimed at GPs and if you even intimated or suggested that GPs did not walk on water, then they just binned it . . . they would never countenance anything that criticised GPs.*

> (Orford, 2018)

Damned

Dammed is a comedy series about social workers and was first shown in 2016. The series stars Jo Brand, whose mother was a social worker and Brand had written the series with Morwenna Banks. In the series she plays Rose Denby alongside Alan Davies, who plays Al and both work in a children's service department, dealing every day with child abuse, youth offending, homelessness and other issues, including the 'tide of bureaucracy and pedantry, and contending with the absurdities and irrationalities of life in a county council office' (British Comedy Guide, 2018). Initial tweeted responses suggested that social work practitioners appear to have received the first series positively e.g. 'hilarious and some very recognisable themes from where I work' (Sophie Olivia 9:11 PM – Sep 27, 2016) and 'I have to say that portrayed the social workers I knew in my time in care, human, funny and beyond all else, caring' (Chris Hoyle 9:32 PM – Sep 27, 2016). Others were less happy with the depiction of social workers e.g. 'Hmmm I don't recognise this at all. For a start I've not seen anyone do any work yet. My team never stop' (Angela Stacey 9:19 PM – Sep 27, 2016). 'I thought it made SW's seem incompetent. I was hoping it would change people's perceptions of us. I'm Disappointed' (Nicola Fone 8:20 AM – Sep 28, 2016).

As an example of the humour used in the show in Series 1 Episode 4 the viewer hears a caller to the social services office say 'This is quite urgent, I am in a restaurant and there is a woman with a baby' the receptionist replies 'ok I am listening' 'she's breastfeeding and it disgusting – you shouldn't allow it.' On IMDb website the highest-rated episode (8.4. out of 10) was the fifth episode in the first series directed by Ian Fitzgibbon, which included an Emergency Strategy Meeting being called by the police officer, who appears to be based in the social work office, as he is convinced two missing Syrian orphans are an ISIS sleeper cell.

In general, *Damned* appeared to receive positive reviews e.g. Anthony Goreham, an IMDb reviewer described the show as 'bittersweet, sympathetic and funny' on 12 November 2016, albeit without 'belly-laughs, it works very well, providing a sympathetic view of an under-appreciated workforce . . . several parts are very funny (the team meeting scenes especially)' (IMDb, 2018).

In 2018 second series of *Damned* was aired and had been positively reviewed e.g. Tim Dowling (2018) was complimentary in his assessment, and argued that it takes 'tremendous skills to find humour in such resolutely unfunny subjects' (Dowling, 2018). Dowling in his comments on the programme asks some important questions:

> *How do you poke fun at social workers without belittling social work? How can you send up absurdly liberal views of sex work without accidentally endorsing absurdly conservative ones? How do you explore the lighter side of the historical sexual abuse of children while still taking it seriously? Damned, thanks to the writers and the cast, has the intelligence and the generosity to allow its characters to be more than one thing, acknowledging the human capacity to embody several contradictory traits: overworked and lazy, spiteful and caring, hopeless and quietly heroic.*
>
> (Dowling, 2018)

In this respect *Damned* provides an example of social work humour that raises the profile of the profession in the media and establishes social work, alongside other professions as a suitable vehicle for situation comedy.

Clare in the Community

Clare in the Community (*Clare*) first started as a cartoon and has also become a radio series. Harry Venning first created *Clare* in cartoon form in 1995 for the *Guardian* newspaper and the title is a play on words relating to *Care in the Community*. The inspiration for *Clare* was Harry Venning's girlfriend at the time who was a social worker. In 2004 the cartoon was adapted for a series for BBC Radio 4 with Harry Venning writing the script with David Ramsden (Elmes, 2009). Clare has a long-suffering partner Brian, who is a teacher and alongside Brian, Clare's colleagues, her sister, and her new neighbours, feature in the cartoon and the episodes. Clare and Brian have a child Thomas Paine Barker who appeared in Series 5 (although does not appear in the cartoons). It's often Clare's lack of self-awareness which drives the humour forward. At times *Clare* has trod a fine line between what could be considered to be offensive, e.g. in an article in 2015 with Alison Benjamin, Malcom Dean the Society editor commented that 'As editor, I have only rejected a couple of strips where I felt Clare was too offensive (albeit unwittingly) to her vulnerable clients. But I heard them when they were recycled for the radio show' (Benjamin, 2015).

In my own research practitioners made reference to *Clare* e.g. [If] 'I'm feeling a bit stressed or just sort of like want cheering up a bit, I'll stick Clare in the Community CDs on, because I, and I find that you would not want any social worker working like that.' And 'A lot of it was about social workers not doing, not practicing what they preached for example . . . I don't think much of the humour around social workers is very supportive of social workers. It's usually to kind of like you know taking the mickey really' (Jordan, 2015). However, there was also a recognition that some of Clare's behaviour, albeit a fictionalised comedy was grounded in and reflected real social work practice:

> *I just find . . . she does, and I think you know says, probably what we have all been tempted to say much the same. She just comes out and says it, . . . and she was gonna give a supervision, but . . . she was dashing out of the door putting her coat on . . . she had been all afternoon this poor student trying to get supervision and when she got it, she was disappearing out of the door with her coat on.*
>
> (Jordan, 2015)

The cartoons fall largely into three main archetypes, first the ones where Clare is practicing as a social worker, either in the office or in her clients' homes (the cartoon uses the word clients as opposed to service user), second her home life with Brian the long-suffering partner and third on wider social/political issues, where *Clare* is located firmly on the left wing of politics and frequently criticises the public spending cuts and the Conservative government.

The cartoons refer to the continued 'battered image of social workers' (Benjamin, 2015) and Clare has been described as 'a selfish, insensitive person and has remained consistently so throughout her cartoon and radio incarnations' (Benjamin, 2015). Harry Venning, *Clare's* creator, has argued that her enduring appeal is the fact that the humour is wider than social work, as it's about 'office politics, and domestic misery' (Benjamin, 2015). Sony Radio Award Panel in 2005 described the radio programme as 'a completely engaging narrative, beautifully observed characters and lots of very funny material.' Venning indicated that he receives mostly positive comments from people who listen to *Clare*, as practitioners comment 'that reminds me of somebody I work with, or . . . . I know somebody in my office like that. I would say its 98% positive [feedback] that I get' (Venning, 2017).

The radio programme was positively reviewed from its inception and retains a large listener audience of between 700,000 and 900,000 listeners and is currently in its twelfth series (Venning, 2017). On the radio the series stars Sally Phillips, Gemma Craven, Alex Lowe and Nina Conti, and one reviewer described *Clare* as a 'joy to watch and listen to, and the script was funny and true to life' (Stoker, 2004). In terms of the response from practitioners, some were concerned at the decline in social work content e.g. 'I've always liked Clare in the Community – both the strip and the radio spin off, even though they're different. But is it just me, or is the actual social-worker content of the strip now almost totally disappeared?' (ElmerPhudd, 22 Jun 2016 7:50). Venning himself acknowledges this, 'In the strip and in the radio show, they don't really do a lot of social work, sometimes we put someone in a room with them, but very rarely they talk about why they are in a room' (Venning, 2017) and for the sake of creating enough material 'she has to be a kind of generic social worker, she has to do children and families and adults, otherwise it would be far too narrow' (Venning, 2017).

The most common type of humour found in *Clare* appears to fit with the superiority theories of humour (humour based on mockery and derision). As a result, we laugh at Clare because we feel ourselves to be morally superior to her, and there are many example of this particularly in the radio series e.g. in Episode 6 Series 1 Clare talks to an Asian husband and wife and says: 'Mr Hadji you've not attended the family centre before, I'll have to take your seat, as I have to sit near the door in case you go berserk and attack me.' Mrs. Hadji then indicates that her husband won't be likely to do that and Clare replies 'Well probably not Mrs. Hadji, but its happened to me more often than you think, once by a colleague.' We find this funny as we feel ourselves superior to Clare insensitivities. Similarly examples can be found in the cartoon and one example published in the *Guardian* on 20 January 2015 shows Clare and Brian drinking in a pub with Megan and a red-haired teacher. The character of a teacher says 'I can't remember when this thought occurred to me but, you know, teachers are social workers too! We also help enable, empower, support, protect, help, reassure and listen.' In response Clare says, 'Maybe it was during one of your twelve weeks holidays?'

Clare is vehemently lefty (even some might argue a *Corbynista*) and frequent targets include the Tory Right, Nigel Farage (particularly complaints about him repeatedly turning up on TV) and Brexit. At different times of its history social work has been more explicitly political in its goals, for example Clement Attlee, was a social worker who became Prime minister and wrote about the social work as a political agitator (Attlee, 2018). Some authors suggest that social work should be taking a more political stance and making allies with politicians who are more closely aligned to social work's orientation to social problems and exposing injustices (Reisch and Jani, 2012; Warner, 2015). Many of the cartoons in *Clare* explicitly make these links with Clare often articulating support for Labour and Jeremy Corbyn. An example of an explicitly political cartoon appeared in the *Guardian* on 8th October 2013: Clare is walking along the street with Megan and says to her: 'I have been a social worker for many years, struggling on in the face of relentless lies, slights, brickbats and bad publicity. But I'm afraid this latest is just too much for the profession to bear.' The punchline appears in the next frame: 'being endorsed by David Cameron at the Tory Party conference!! How are we going to live that down?' This appears to be one of the most popular cartoons and was shared 136 times, much more than other cartoons that year.

Conclusion

This chapter has not considered serious depictions of social workers, so what has been the value in considering humourous representations? In my experience the social work profession has tended to complain about its depiction, or more often its lack of exposure in the popular media, so might there be a value to the profession of being depicted, even if it's in a humourous way? What could the implications be for practice? If according to the Lacanian one-liner 'the truth has the structure of a fiction' (Žižek and Mortensen, 2014, p. 41) might these humourous and fictional tales of social workers hold some truths in them about social work practice?

Many of the comedic portrayals of social workers make errors in the way social work is practiced, for example frequently they show social workers taking action without any recourse to court, so should the profession be concerned with blatantly unrepresentative portrayals and their impact on practice? Yes, in the sense that if vulnerable people see social workers behaving in oppressive and unfair ways then this does have an influence even if it is superficial and short term. Research by Franklin (1998) identified two often repeated media stereotypes of social workers: 'woolly minded, indecisive, ineffectual, incompetent wimps, or authoritarian, bullying, bureaucrats, who speak in a jargon and are engaged in legalised baby snatching' (Franklin, 1998) and the cartoon depictions of social workers reinforce that finding.

The mocking depictions of social workers in cartoons aimed primarily at children such as *King of the Hill* and the *Simpsons* raise the question of the usefulness in terms of practice of such portrayals to a profession which arguably has limited exposure. Certainly if this is the first version of social workers that children will be exposed to, the depiction will be likely to have negative impacts, if any, as the characters often appear as bit-part players to the main narratives. They are also more likely to be portrayed as male which is disconcerting and unrealistic for a profession which is predominantly female. This seems to continue in the televised portrayals and even in *Damned*, which is written by two women, nearly half of the central characters are men. Fran's cartoons often feature white men and he gave a clue to why this might be 'I discovered quite early on . . . the people you can take the mickey out of, without anybody getting worried about it, is of course middle-aged white men – nobody's going to get offended about that' (Orford, 2018).

Responses to the rendering of social workers in the media, particularly in on-line comments, suggest a level of insecurity from practitioners about social work's humourous presentations in the popular media. However these often do not go unchallenged e.g. in response to Wright's (2011) comments about *South Park*'s portrayal of social workers one respondent said

> *You're concerned that the habitually irreverent and purposefully controversial show South Park is giving the title of 'Social Worker' a bad name? I am far more concerned about your hypersensitivity and lack of sense of humor giving those involved with the social work profession a bad name. If we can't laugh at ourselves (even blatantly unrepresentative caricatures of ourselves) I fear for as a profession and the individuals we are charged with helping.*

So whilst practitioners and others who care about the profession should not ignore the media presentations of social work, perhaps social work can be generous in our attitude to the obvious and sometimes crude caricatures of what social work is and be thankful that at least social work is receiving some attention, rather than none. Often comedy shows are parodies and what is real life and we should not attach too much weight to such depictions. At the very least it starts a debate about what social work is and what social workers do, particularly with children who may see cartoons such as the *Simpsons*.

Could it be that the inclusion of social workers as ordinary characters would be much more inclusive than the occasional portrayal of a social worker as the bit part figure, so that the depiction of a person who happens to be a social worker, while they could be the butt of some jokes, are also there as part of the makeup of the show? This, one could suggest, is what *Damned* has attempted to do. Some reviewers found the portrayal in *Damned* to be sympathetic to an underappreciated workforce and this seems worthy in itself.

This could also be true for *Clare in the Community*, but do social workers and the social work profession fear *Clare* because she is like us in ways we do not like, and she shows social workers at their worst? Is the characteristic she portrays so obviously untrue that at least she is a social worker and can be seen to have placed social work at least in the consciousness of some, albeit if 'they are only radio 4 listeners or Guardian readers,' we should be thankful that someone is making social work the main character?

Clare also represents something about social work that we have lost, she is unapologetically generic in her approach and this might ironically be a future model for a more engaged practice of social work, attached to small patch based teams, a future where social workers engage, like Clare, with the whole community, with whoever needs help. I conclude this section with the comments from Sally Phillips who plays Clare in the radio programme. In an interview with Lisa Higgins (2016) she championed the role of social workers: 'I've been the voice of the radio incarnation of Clare in the Community for 10 years now. I love Clare . . . I have social workers in my friendship group and a niece who is a paediatric social worker. It's tough stuff' (Phillips, 2016).

References

Attlee, C. R. (2018) *The Social Worker* London: Forgotten Books (original London: G. Bell and Sons Ltd. 1920).

Benjamin, A. (2015) Clare in the community: The first 20 years *The Guardian* November 24 www.theguardian.com/society/2015/nov/24/clare-in-the-community-20-years-exhibition

Bradley, L. (2016) 20 seasons in, Matt Stone and Trey Parker reveal the secret to keeping South Park cool *Vanity Fair* www.vanityfair.com/hollywood/2016/09/south-park-20th-anniversary-interview (accessed 12/3/18).

British Comedy Guide (2018) *Damned* www.comedy.co.uk/tv/damned/ (accessed 24/7/18).

Byrne, C. (2004) BBC viewing figures fall to all-time low *The Independent* www.independent.co.uk/news/media/bbc-viewing-figures-fall-to-all-time-low-754162.html (accessed 6/3/18).

Chemers, M. and Karamanos, H. (2012) Chapter 3 'I'm not special' Timmy, Jimmy and the double-move of disability parody in South Park in Cogan, B. (Ed.), *Deconstructing South Park: Critical Examinations of Animated Transgression* Plymouth: Lexington Books.

Cogan, B. (Ed.) (2012) *Deconstructing South Park: Critical Examinations of Animated Transgression* Plymouth: Lexington Books https://ebookcentral.proquest.com/lib/universityofessex-ebooks/detail.action?docID=817165 (accessed 12/3/18).

Crace, J. (2017) Priti Patel bites the dust-after leaving a hell of a carbon footprint Ignore me please (but do come to my party) John Crace's digested week *The Guardian* Saturday November 10.

Delingpole, J. (2010) What 'The Simpsons' can teach us about life *The Telegraph* www.telegraph.co.uk/culture/tvandradio/6872394/What-The-Simpsons-can-teach-us-about-life.html (accessed 6/3/18).

Dessau, B. (2012) *Beyond a Joke: Inside the Dark Minds of Stand-Up Comedians* London: Arrow.

Dowling, T. (2018) Damned review: Pitch black comedy finds the funny in social services *The Guardian* February 14 (accessed 6/3/18).

Eborall, C. and Garmeson, K. (2001) *Desk Research on Recruitment and Retention in Social Care and Social Work* London: Business and Industrial Market of Research.

Elmes, S. (2009) *And Now on Radio 4: A Celebration of the World's Best Radio Station* London: Random House.

Fast Company (2009) *The Simpsons, by the Numbers* www.fastcompany.com/1460994/simpsons-numbers (accessed 6/3/18).

Franklin, B. (1998) *Hard pressed: National newspaper reporting of social work and social services* Sheffield: Report for University of Sheffield.

Hooton, C. (2015) Family Guy is moving from BBC3 to ITV2 *The Independent* www.independent.co.uk/arts-entertainment/tv/news/family-guy-is-moving-from-bbc-three-to-itv2-10128106.html (accessed 12/3/18).

Internet Movie Database (2016) *Damned Episode 5 Series 1* www.imdb.com/title/tt5865528/?ref_=tt_eps_rhs_0 (accessed 12/3/18).

Internet Movie Database (2018) www.imdb.com/title/tt0576956/ (accessed 12/3/18).

Jordan, S. (2015) *That joke isn't funny anymore: Humour, jokes and their relationship to social work.* London: University of East London: Professional doctorate thesis.

Legood, A., McGrath, M., Searle, R. and Lee, A. (2016) Exploring how social workers experience and cope with public perception of their profession *British Journal of Social Work* 46, 1872–1889 (accessed 13/1/16).

Martin, R. (2010) The Simpsons Season 13 DVD Review *411Mania* https://411mania.com/movies/the-simpsons-season-13-dvd-review/ (accessed 6/3/18).

Miller, A. and Bartlett, S. (Eds.) (2004) *The Changing Face of Social Care: How Social Care Has Evolved through the Ages and How It Can Re-Invent Itself for the Recruitment Challenges Ahead* London: Community Care.

Orford, F. (2017) *Social Work Cartoons* www.francartoons.co.uk/wp/social-work-cartoons/

Orford, F. (2018) Interview with the author 12th April 2018.

Perry, N. (2005) *Getting the Right Trainers What Enables Social Workers to Do a Good Job?* Oldham: ATD Fourth World Ltd.

Phillips, S. (2016) Why I'm cheering from the side-lines for care workers *The Guardian* March 8 www.theguardian.com/society/2016/mar/08/sally-phillips-cheering-care-workers-care-awards (accessed 24/7/18).

Power, E. (2015) Family Guy: The Simpsons Guy, review: 'The humour was forced throughout' *The Telegraph* www.telegraph.co.uk/culture/tvandradio/tv-and-radio-reviews/11710388/Family-Guy-The-Simpsons-Guy-review-the-humour-was-forced-throughout.html (accessed 12/3/18).

Reisch, M. and Jani, J. S. (2012) The new politics of social work practice: Understanding context to promote change *British Journal of Social Work* 42, 1132–1150. DOI: 10.1093/bjsw/bcs072

Shattuck, K. (2009) It was good to be 'king,' but what now? *New York Times* April 22 www.nytimes.com/2009/04/26/arts/television/26shat.html (accessed 6/3/18).

Stevenson, L. (2017) *BASW Writes to BBC over 'Cardboard Cut-Out' Social Worker Portrayal in EastEnders* October 20 www.communitycare.co.uk/2017/10/20/basw-writes-bbc-cardboard-cut-social-worker-portrayal-eastenders/

Stoker, G. (2004) *Clare in the Community BBC Radio 4 Live Recording the Drill Hall* www.britishtheatreguide.info/reviews/clarecommunity-rev

Strang, D. (2003) *Keynote Speech to ADSW Annual Conference*, Thursday 22 May, Dunblane.

Tower, K. (2000) In our own image: Shaping attitudes about social work through television production *Journal of Social Work Education* 36(3), Fall, 575–585. Published by: Taylor & Francis, Ltd. on behalf of Council on Social Work Education Stable URL www.jstor.org/stable/23043531 (accessed 5/3/18 16:15 UTC).

Venning, H. (2017) Interview with the author 14 December 2017.

Wain, D. and Stevenson, L. (2016) 10 of the best and worst TV social workers Community Care looks at 10 social workers on TV and lists the varying qualities *Community Care, Social Workers on TV* June 6 www.communitycare.co.uk/2016/06/06/10-best-worst-tv-social-workers/

Warner, J. (2015) *The Emotional Politics of Social Work and Child Protection* Bristol: Policy Press.
Wright, G. (2011) Latest 'South Park' episode raises concern about social work title *Social Worker Speak* www.socialworkersspeak.org/hollywood-connection/latest-south-park-episode-raises-concern-about-social-work-title.html (accessed 12/3/18).
Žižek, S. and Mortensen, A. (2014) *Žižek's Jokes: (Did You Hear the One about Hegel and Negation?)* London: MIT Press https://ebookcentral.proquest.com/lib/universityofessex-ebooks/detail.action?docID=3339752 (accessed 12/3/18).

5 'Deep down'

Humour, health and relationship-based practice

Q: Why do they bury Social Workers 300 feet in the ground?
A: Deep down they are really good people.

Introduction

This chapter considers the positive aspects of humour and how it can facilitate and help in relationships. As seen in Chapter 3 using humour can help social workers cope with the stress of the work, and if used selectively, sensitively and appropriately, humour can be seen alongside other mechanisms as a positive method of dealing with stress. Developmental theories of humour, see humour as important for a child's growth and attachment, as humour creates the bonds between caregiver and child, as it demonstrates pleasure in each other's company. This can also apply in social work, in that social workers attach to teams in which they find warm reciprocal bonds and by extension this experience enables them to make an attachment to the profession of social work. If social workers feel themselves insecure and unattached they will be less likely to remain in teams or in the profession and humour use can provide social workers with reason to positively bond with their teams. This chapter examines the research claims for this, includes practice examples and the links to attachment and health.

Humour, health and mental well-being

Whilst laughter cannot automatically be associated with humour, there have been a number of studies which link humourous laughter to beneficial effects and specifically in terms of an individual's health. Laughter has long been viewed as 'good medicine' for a variety of illnesses (Bennett and Lengacher, 2006). Coser (1959) found that some patients 'taught' other patients through jocular interactions to cope with the hospital environment. It has long been part of folk mythology that being happy makes you feel good. '*A cheerful heart is good medicine, but a crushed spirit dries up the bones*' (Proverbs 17:22, *NIV*).

The notion that psychosocial considerations can help to maintain good mental health was put forward by Cousins in the 1970s as he speculated that if stress could worsen his condition, then feeling good could improve his health, and as result put himself on a self-medicating regime of humorous videos, which put his disease into remission (Bennett and Lengacher, 2006).

Other research suggests that humour and laughter helps individuals cope more effectively with stress by enhancing greater social support (Martin, 2002) and reduces muscle tension (Roscoe, 2017). Bennett et al. (2014) found that laughter and humour had positive applications in the treatment of kidney disorders and dialysis as it decreased anxiety in patients. It has even been argued that deploying humour in public health advertising can have positive effects in promoting public health issues and can be an effective way to change attitudes in high-anxiety settings, such as testicular cancer examinations, or times when patients are nude, and in safe sex discussions with adolescents (Yoon, 2015). Schneider et al. (2018) found that it was specific styles of humour which are associated with positive mental health, and they found aggressive styles of humour or self-defeating humour were more likely to be correlated with poorer mental health and that understanding such humour use would help therapists. Studies also show positive benefits of humour use amongst older people which mitigate against the negative effects of ageing 'it is difficult to determine which cognitive and/or emotion regulatory mechanisms are operative . . . [and] little is known about the overlap across the person/social factors reviewed – social support, self-efficacy, spirituality, and humour' (Marziali et al., 2007, p. 7).

There is an abundance of research and studies espousing the benefits of humour and laughter for health (Longo, 2010), although the evidence is less convincing than media reports sometimes promote. For example, Martin (2002) found that there was limited empirical evidence for the many beneficial claims of laughter on immunity, pain tolerance, blood pressure and concluded that the enthusiastic and wide reaching claims for humour such as its relationship to 'longevity, recovery from illness or production of endorphins do not withstand close analysis' (Martin, 2002, p. 218).

So, whilst there are significant claims made for humour and laughter in relation to health, few studies explain why humour should have such a positive impact on an individual's health, and in the next section considers this in the context of relationship-based social work practice.

Relationship-based social work (RBSW) practice

In considering social work in relation to health there has been a strong tradition in social work of psychodynamically informed practice. This can be dated back to Florence Hollis' classic text on casework, which combined psychoanalysis and social work theory (Cooper, 2018). Not only in social work but in nursing too (e.g. Hedges et al., 2012) there has been growing body of practitioners who described their practice as relationship based.

In social work the tradition of psychodynamically informed practice was particularly influential between 1930 and 1970 (Howe, 1998) and Howe (1998) argued that if poor relationships are where psychosocial competences go awry, then good relationships are where they are likely to recover. For some commentators the relationship between a social worker and the child or family with whom they are working is both a means of intervention but also the *intervention* itself (Wilson et al., 2011). A number of writers have outlined what elements constitute good relationship-based practice e.g. Ruch (2005, 2012) focussed on reflection and suggested that relationships require openness and willingness to 'commit something of oneself' (Ruch et al., 2010, p. 245). Murphy et al. (2013) suggested a close connection between relationship-based practice and person-centred approaches in social work. Ruch et al. (2010) pointed out that 'good relationships take commitment, hard work and imagination and when they work they offer a vulnerable or emotionally damaged person the possibility of encountering themselves in a new and positive way-worthy of another's interest and respect' (Ruch et al., 2010, p. 246). However neither Howe (1998), Trevithick (2003), O'Leary et al. (2013) or Wilson et al. (2011) made mention of humour use in their work on relationship-based contemporary practice in social work and so humour, an omnipresent component of social life, has received little attention in regard to relationship-based social work practice.

Kadushin and Kadushin (1997) and Hill and O'Brien (2004) have warned about the use humour in relationships and point out that humour and sarcasm should never be used at the service users expense, as this conveys a lack of empathy and insensitivity. However, given the ubiquitousness of humour in all human relationships, humour is something that cannot be avoided, and instead social workers have to engage with humour, either as the object of humour or as active participants.

Contemporary trends in social work practice suggest that in the 1990s there was a move away from psychodynamically or psychosocially informed practice and towards a more regulated, managerialist, procedural and legalistic culture in social work (Wilson et al., 2011). Some have linked the move away from relationship-based practice with the prioritising of the care management approach: 'The research suggests that the effective childcare professional remains an enigma. The current emphasis given both to prioritising high risk cases and to a care management approach concerned with assessment, organizing provision, monitoring and review, threaten to combine to minimise, if not eliminate, what has been the core social work task of working for personal and social change . . . through insights gained form the social work relationship' (Leigh and Miller, 2004 pp. 263–264).

This created a concern that with an increased focus on procedures social work practice had lost its focus on relationships, and a more procedurally based practice became less humane and less effective for social workers and their service users.

Social work is a risky endeavour, fraught with contradictions, anxiety and complexities, and the practice of social work one could argue is primarily about risk taking. An important question arose as to why social workers would take the risk of saying something which could be deemed to be at best inappropriate and at worst against the standards of proficiency, and possibly result in them losing not only their job, but ending their career? What could be the value to a social worker in using humour and taking such a risky course of action? In terms of managing risky situations the answer could lie in part in what humour communicates about the teller to the recipient, as humour as a universal human characteristic conveys 'normality' to others and communicates their humanity, because it is founded in our earliest attachment experiences (Nelson, 2012). In this sense humour has unique power to convey a particular characteristic about a social worker, and that is why social workers take the risk of using humour, as the opposite, a lack of humour, conveys a lack of humanity.

Social workers can often want to be seen as humourous as the opposite, 'humourless individuals,' are valued less by their service users and colleagues. In the words of Stephen Leacock the essence of humour is human kindliness and the opposite of humour is deathliness (Lynch, 1988). In this respect humour conveys a useful and important message to the recipient, that the humour or joke teller shares a common humanity with the recipient, which can form the basis of the social work relationship.

As Frost (1992) argues the use of humour with service users is always risky, but he suggests that it has a place in helping relationships. Service users can themselves teach social workers the importance of finding the humour, irony and absurdity in their situations, and whilst it is unethical to laugh at people and their problems, it may be helpful to laugh with them as they describe the humourous aspects of their experiences (Frost, 1992). This I would describe as 'laughing in a helpful way' and even individuals who have lived through some of the most barbaric extremities of human behaviour appreciate the value of humour (Frankl, 1946).

Psychosocial approaches and attachment

Psychotherapeutic and psychosocial approaches to working with people have had a long tradition of engaging with humour and jokes as ways of working with individuals in distress (Richman, 1996; Fabian, 2002). Arguably this was born out of Freud's classic work (1960) which gave rise to a body of literature which saw jokes and humour as legitimate ways into understanding unconscious behaviour and actions.

People do not react to humour in the same way and humour can be used as an excuse for unacceptable behaviour or in contrast a positive way of managing stress (Moran and Hughes, 2006). As Lemma (2000) has pointed out when people communicate using humour, they often have more than one meaning behind the humour 'it's funny, but it's serious, deadly serious sometimes' (Lemma, 2000, p. 21).

Our earliest relationship starts in humour and some argue that humour is central to human development (Bateson, 1953). Bateson (1953) argued that humour is an evolutionary step in the human species and found that the ability to discriminate between messages encoded at different levels of abstraction was inherent in the development of playful activities, humour amongst them. As indicated earlier evolutionary scientists suggest that humour has played a vital role in the development of the unique intellectual and perceptual abilities of humans (Clarke, 2008). Other developmental theorists have focussed on language development in children and researchers have highlighted how developing children particularly enjoy jokes and riddles (King-DeBaun, 1997; Musselwhite and Burkhart, 2002), suggesting that humour has a key developmental aspect.

As noted in Chapter 2 it is possible that humour could be located in our earliest development. Bowlby (1999) argued that smiling was crucial to developing attachments and Nelson (2012) noted that laughter is a process for the caregiver and babies to attune to each other and is the beginning stage of how we learn to safely interact with one another. Laughter and humour therefore provide a secure base for exploring the world, and Nelson (2012) makes the case for considering laughter as a form of attachment behaviour, as important if not more so than crying. This theme has a 'track record' in psychosocial thought and Bollas (1995) suggested that a sense of humour is essential to human survival, and that the mother who develops their babies' sense of humour is assisting her to detach from 'dire mere existence' and that as an adult, they will find humour in the most awful circumstances, ultimately benefiting from the origins of the comic sense.

Practice issues

Humour use has positive benefits for individuals' physical and mental health and this is no less true for social workers, than it is for everyone. Social workers have to negotiate complex and difficult aspect of relationships and social workers have used humour in working with people with dementia, individuals in crisis, people dealing with illness and even in child abuse (Moran and Hughes, 2006). What links these practices together and how effective were such practices in helping people? As De Boer and Coady (2007) have argued, it is the relationship which become so important for services users. Social workers interviewed as part of my research into humour use (Jordan, 2015) indicated that humour was central to social work practice and important for creating and maintaining work relationships:

> I noticed myself that it seems to be the social workers who can laugh off cases, and you know have a quip here or there, an anecdote available at the drop of a hat, ... they ... seem to belong to groups or cliques and it appears that they are readily kind of recruited into, you know the group that goes down to the pub at the end of the week.
>
> (Jordan, 2016)

The comment that social workers need to 'laugh off cases' is a reference to managing some of the anxieties and emotional impact the work can have on social workers. However, it is important to note that there is an element of exclusion, revealed in this comment, as this social worker feels excluded from these groups, unhappy that he is not invited or recruited to the group which goes to the pub. Therefore, whilst humour may help some social workers to make friendships and engage with others, there is a contradictory aspect of humour, as at the same time it connects and engages people, it also excludes others who are not included in the humour.

Another social worker made a comment which illustrates the important role shared experiences have in creating group solidarity at work. In response to my question about how humour effects relationships at work Zia replied:

> *All the things which sprung to mind, were all the things that were half funny and half horrendous. When you get the context it's quite erm difficult because you can tell things aren't funny in so many ways as well and only . . . other workers get that . . . it's so peculiar, the work a lot of the time . . . to work in this world where everything is different to your normal life in so many ways . . . so the people you work with are damaged or so troubled quite often, that you could not have a friendship with them or you would not even sustain it . . . I feel like I live in two worlds – there is my world and then there is the people I work with.*

(Jordan, 2016)

Zia was struggling to reconcile the emotional experiences of living and working in different worlds, and Zia is making an important point about the place of social work in her mind and the external reality of the work. In discussing the humour in her team Zia commented that 'funny situations' were 'half funny and half horrendous,' and suggested that only social workers could understand why this would be so. She indicated that as social workers share similar experiences, they can find humour in such situations, which would not be seen as funny or humorous by people 'outside of social work,' and in this respect humour can create group solidarity, through a shared experience of this work life. However, this can also create tensions between those outside social work, who do not share these experiences.

Social workers practice in a world which places contradictory demands on them; Zia's experiences are not uncommon for many social workers, and living in two worlds is a theme which other writers have articulated (e.g. Waddell, 1985; Becker and MacPherson, 1988). In the extract above Zia spoke about how the 'horrendous' world of child abuse contrasted with her own life and personal experiences. This is important as it illustrates that social workers use humour to negotiate and manage the emotional contradictions the work raises for them.

Humour is common to human experiences and it is possible that humour has a unique potential for demonstrating particular characteristics of a social worker, so when applied sensitively and appropriately it could be a useful tool to enable social workers to help service users manage their own emotions:

> *Actually* [humour] *can be really quiet a powerful tool in relieving people's sense of anxiety, particularly if they are talking to a social worker and they have had ideas in their head that social workers can only be quite a punitive thing through children's services . . . I think it is very important to get over that, that boundary, and I think any humour, anything which makes you seem more human, nothing really does that as well as humour.*

(Jordan, 2016)

The effect of humour applied by this social worker is then to distance themselves from others social workers, and also to make himself appear 'more human.' It may be that this interviewee wanted to unconsciously communicate that he is different from 'punitive' social workers, and in this way the humour is used in a way to make the social worker seem less threatening and more 'humane.'

Social workers used humour to make a joke about their own behaviour, for example Jerry commented:

> *I say to people you know . . . maybe at the end of a working relationship I would say things like: I mean this in the nicest way, but I hope we never see each other again type of thing . . . if I'm working out a plan with a family, . . . I will have a little bit of a joke with them and say 'you're going to get fed up with the sight of my face' . . . I do try to lighten up a situation . . . that's potentially a serious situation.*

(Jordan, 2016)

The comment illustrates the use of humour as a way of defusing the emotions and anxiety service user's face when working with a social worker. This suggests that it is humour, which allows the social worker to 'appear more approachable' and to be a safe mechanism for acknowledging the anxiety a service user may be feeling. It illustrates in my view the theme that humour operates in social interactions, to play a unique role in making the individuals which use humour appear to be 'more human.' One could suggest here that this may also be the unconscious anxieties of the worker, as they try to defend against their own feelings of anxiety, the encounter poses for themselves. Humour is crucial to effective social work practice:

> *I would just be worried about sending someone out to work with people if they could not have a bit of a laugh or to laugh at themselves – not inappropriately, sometimes what we are doing isn't . . . a bit funny but I kind of think that if they don't have [a sense of humour] . . . then they are really not up to the job, it's about working with people.*

(Jordan, 2016)

This suggests that not only do social workers need to have sense of humour to be able to effectively carry out the work, it would be concerning, at least to this team manager, if a social worker did not have the ability to laugh at themselves. What this suggests is that a social worker's ability to laugh at oneself is a beneficial characteristic in creating relationships with others. This appears to be in line with a reflective approach in social work, where the ability to reflect back on a social worker's own behaviour and process it in a way which is 'emotionally safe,' an ability to laugh at oneself is critical to effective practice and managing one's emotions:

> *Sometimes, for instance, I would be working with a mum and child and there is another toddler in the room who is busy playing with the toys in the*

corner and will do something just randomly quite funny, enough to generate
a laugh – I can share that humour, that can be quite a good leveler.

(Jordan, 2016)

Therefore showing a 'playful and showing a human side' of the social worker is crucial for establishing and maintaining relationships with others. There is a need for play and playfulness in human engagements, particularly with children, in order to effectively communicate and form relationships with them, particularly in the development of the self (Nicolson, 2014). It would be difficult to imagine that a child would feel comfortable talking to a social worker who presented as unable to operate in this way. In this sense humour conveys a particular subtle and unconscious message to the recipient, that the user of humour is '*worth making a relationship with.*'

Building a rapport and establishing relationship with adults and children requires time, sensitivity and patience. Humour, this social worker suggests, can be a way of measuring the establishment of a relationship with an adult or a child. In the following extract Quinn gave an example of how this can 'normalise' a relationship with the parent of an autistic child:

I remember one parent telling me – she said to her [autistic] son 'her hair
needs cutting badly' [he asked] 'why would you want it cut badly?' you know
she said it to me as a joke, but if I say that someone, they will go to laugh and
think 'oh, should I be laughing or not?' . . . I think joking with someone is a
good indication of the strength of the relationship, and that they have, if you
have built up good relationships.

(Jordan, 2016)

Humour therefore can become a measure of how comfortable parents and children are with the social worker, but the worker also acknowledges that people both inside and outside of social work, might be anxious that such interaction might risk being seen as inappropriate.

Conclusion

Humour is necessary for successful relationships, but is also a contradictory, risky undertaking and practitioners frequently engage with humour as its key to social life. Attachments are also crucial to establishing and maintaining relationships and from our earliest development laughter and humour helps parents and children bond, and in later life creates and maintains relationships with others. At the same time, it helps social workers manage their own emotions and the emotions of others. Social workers fear not being taken seriously, but conversely the use of humour and sharing of jokes can help social workers manage their unhappiness at work, to cope with the stress of the work, and to have successful relationships.

Social workers practice in a world which places contradictory demands on them (Evans and Harris, 2004). Indeed it might be suggested that the role of social work is inherently contradictory. My findings suggest that humour enables social workers to remain mentally healthy and manage those contradictions and to hold conflicting views at the same time. Existing in 'two worlds' is a theme which other writers have written about (e.g. Waddell, 1985; Becker and MacPherson, 1988). As social workers work in sharply 'contradictory worlds,' social workers have to find ways of reconciling these worlds and part of this negotiation involves tolerating the contradictory nature of the work. Social workers are in Ungar's (2004) phrase the 'inside outsiders,' working both inside homes and outside social norms and conventions, operating across the two worlds of the social worker and the service user, the public policy and private pain. Social workers are both a part of society (and the state) and apart from society (working alongside an often socially marginalised minority).

Humour enables social workers to tolerate and negotiate these contradictory worlds. Whilst humour may help some social workers to make friendships and have good working relationships, the first contradictory aspect of humour, is that at the same time it connects and engages people, it also excludes others who are not included in the humour. At times humour works to create solidarity amongst social workers, particularly in the face of perceived injustices by management, or frustration with the behaviour of service users. In this sense social workers create solidarity amongst themselves at the price of excluding others. This parallels Carden's (2003) theory whereby social workers use humour as a vehicle for 'maximum consciousness' through which the culture of social work can articulate its own distress and occasions of reciprocated humour can create relationships with co-workers and service users and reduce hierarchical differences (Robert and Wilbanks, 2012).

This need to reconcile contradictory or uncomfortable feelings about the work, can result in cognitively separating and not thinking. This process of 'psychic economy' is often referred to as splitting, which can lead to polar simplification and classification (Freud, 1964). The use of humour by social workers is an enabling process whereby they manage contradictory feelings about their management, service users and even themselves. Social workers avoid the process of splitting by using humour to reconcile contradictory feelings, because humour allows two separate and contradictory views about something to be held at the same time and these may well be the 'emotional regulatory mechanisms' that create social support which Marziali et al. (2007) refer to, and it is likely that the measure of a 'healthy relationship' is one which is exemplified by the use of pro-social humour in relationships.

References

Bateson, G. (1953) The position of humour in human communication in Foerster, H. (Ed.) *Cybernetics* New York: Josiah Macy, Jr. Foundation.

Becker, S. and MacPherson, S. (Eds.) (1988) *Public Issues, Private Pain: Poverty, Social Work and Social Policy* London: Care Matters.

Bennett, M. P. and Lengacher, C. A. (2006) Humour and laughter may influence health II: Complementary therapies and humour in a clinical population *Evidence-based Complementary and Alternative Medicine* 3, 187–190.

Bennett, P. N., Parsons, T., Ben-Moshe, R., Weinberg, M., Neal, M., Gilbert, K., Rawson, H., Ockerby, C., Finlay, P. and Hutchinson, A. (2014) Laughter and humor therapy in dialysis *Seminars in Dialysis* 27(5), September–October, 488–493. DOI: 10.1111/sdi.12194

Bollas, C. (1995) *The Work of Unconscious Experience* London: Routledge.

Bowlby, J. (1999) *Attachment and Loss* (Vol. 1, 2nd ed.) New York: Basic Books.

Carden, I. (2003) On humour and pathology: The role of paradox and absurdity for ideological survival *Anthropology & Medicine* 10(1), 115–142.

Clarke, A. (2008) The science of laughter *The Times* September 13.

Cooper, A. (2018) *Conjunctions: Social Work, Psychoanalysis and Society* London: Karnac.

Coser, R. L. (1959) Some social functions of laughter: A study of humour in a hospital *Human Relations* 12, 171–182.

De Boer, C. and Coady, N. (2007) Good helping relationships in child welfare: Learning from stories of success *Child and Family Social Work* 12, 32–42.

Evans, T and Harris, J. (2004) Street-level bureaucracy, social work and the (Exaggerated) death of discretion *British Journal of Social Work* 34, 871–895.

Fabian, E. (2002) On the differentiated use of humour and joke in psychotherapy *The Psychoanalytic Review* 89(3), June.

Frankl, V. (1946) *Man's Search for Meaning* London: Hodder and Stoughton.

Freud, S. (1960) *Jokes and Their Relation to the Unconscious* London: Routledge & Kegan Paul.

Freud, S. (1964) *Splitting of the Ego in the Process of Defence* London: Hogarth Press.

Frost, C. (1992) *Having Fun in Social Work* Middle Tennessee State University Paper http://capone.mtsu.edu/cfrost/soc/thera/HUMOR.htm

Hedges, C. C., Nichols, A. and Filoteo, L. (2012) Relationship-based nursing practice: Transitioning to a new care delivery model in maternity units *Journal of Perinatal and Neonatal Nursing* 26(1), January–March, 27–36. DOI: 10.1097/JPN.0b013e31823f0284

Hill, C. E. and O'Brien, K. M. (2004) *Helping Skills: Facilitating Exploration, Insight and Action* Washington: American Psychological Association.

Howe, D. (1998) Relationship-based thinking and practice in social work *Journal of Social Work Practice: Psychotherapeutic Approaches in Health, Welfare and the Community* 12(1), 45–56. DOI: 10.1080/02650539808415131

Jordan, S. (2015) *That joke isn't funny anymore: Humour, jokes and their relationship to social work.* London: University of East London. Professional doctorate thesis.

Jordan, S. (2016) Relationship based social work practice: The case for considering the centrality of humour in creating and maintaining relationships, *Journal of Social Work Practice.* DOI: 10.1080/02650533.2016.1189405

Kadushin, A. and Kadushin, G. (1997) *The Social Work Interview* New York: Columbia University Press.

King-DeBaun, P. (1997) *Computer fun and adapted play: Strategies for cognitively or chronologically young children with disabilities part 1 and 2.* Proceedings of Technology and Persons with Disabilities Conference, California University, USA.

Leigh, S. and Miller, C. (2004) Is the third way the best way: Social work intervention with children and families *Journal of Social Work* 4(3), December, 245–267.

Lemma, A. (2000) *Humour on the Couch* London: Whurr.

Longo, M. (2010) Humour use and knowledge-making at the margins: Serious lessons for social work practice *Canadian Social Work Review/Revue canadienne de service social* 27(1), 113–126.

Lynch, G. (1988) *Stephen Leacock: Humour and Humanity* Quebec, Canada: McGill Queens University Press.

Martin, R. A. (2002) Is laughter the best medicine? Humor, laughter, and physical health current *Directions in Psychological Science* 11(6), December, 216–220.

Marziali, E., McDonald, L. and Donahue, P. (2007) *The role of coping humour in the physical and mental health of older adults* Social and Economic Dimensions of an Aging Population Research Papers 225, McMaster University.

Moran, C. C. and Hughes, L. P. (2006) Coping with stress: Social work students and humour *Social Work Education* 25(5), August, 501–517.

Murphy, D., Duggan, M. and Joseph, S. (2013) Relationship-based social work and its compatibility with the person-centred approach: Principled versus instrumental perspectives *British Journal of Social Work* 43, 703–719.

Musselwhite, C. and Burkhart, J. (2002) *Social scripts: Co-planned sequenced scripts for AAC users*. Proceedings of Technology and Persons with Disabilities Conference, California State University, CA, USA.

Nelson, J. K. (2012) *What Made Freud Laugh: An Attachment Perspective on Laughter* London: Routledge.

Nicolson, P. (2014) *A Critical Approach to Human Growth and Development* Basingstoke: Palgrave Macmillan.

O'Leary, P., Tsui, M.-S. and Ruch, G. (2013) The boundaries of the social work relationship revisited: Towards a connected, inclusive and dynamic conceptualisation *British Journal of Social Work* 43, 135–153.

Richman, J. (1996) Points of correspondence between humor and psychotherapy *Psychotherapy* 33(4), Winter.

Robert, C and Wilbanks, JE (2012) *The Wheel Model of humor: Humour events and affect in organizations* Human Relations 2012 65: 1071 originally published online 27 April 2012 DOI: 10.1177/0018726711433133

Roscoe, L. A. (2017) Sometimes laugher is the best medicine *Health Communication* 32(11), 1438–1440. DOI: 10.1080/10410236.2016.1227295

Ruch, G. (2005) Relationship-based practice and reflective practice: Holistic approaches to contemporary child care social work *Child and Family Social Work* 10, 111–123.

Ruch, G., Turney, D. and Ward, A. (Eds.) (2010) *Relationship-Based Social Work: Getting to the Heart of Practice* London: Jessica Kingsley.

Ruch, G. (2012) Where have all the feelings gone? Developing reflective and relationship-based management in child-care social work *British Journal of Social Work* 42, 1315–1332.

Schneider, M., Voracek, M. and Tran, U. S. (2018) 'A joke a day keeps the doctor away?' Meta-analytical evidence of differential associations of habitual humor styles with mental health *Scandinavian Journal of Psychology* February 12. Date of Electronic Publication February 12. DOI: 10.1111/sjop.12432

Trevithick, P. (2003) Effective relationship-based practice: A theoretical exploration *Journal of Social Work Practice: Psychotherapeutic Approaches in Health, Welfare and the Community* 17(2), 163–176. DOI: 10.1080/026505302000145699

Ungar, M. (2004) *Nurturing Hidden Resilience in Troubled Youth*Toronto: University of Toronto Press.

Waddell, M. (1985) Living in two worlds: Psychodynamic theory and social work practice *Free Associations* 10.

Wilson, K., Ruch, G., Lymbery, M. and Cooper, A. (2011) *Social Work an Introduction to Contemporary Practice* Harlow: Pearson Education Ltd.

Yoon, H. J. (2015) Humor effects in shame-inducing health issue advertising: The moderating effects of fear of negative evaluation *Journal of Advertising* 44(2), 126–139. Copyright 2015, American Academy of Advertising ISSN: 0091-3367 print / 1557-7805 online. DOI: 10.1080/00913367.2015.1018463 Southern Methodist University, Dallas, TX, USA.

6 'The most PC people around'

Social work, humour and political correctness

A senior practitioner was in the team room and overheard the team manager giving advice to two social workers who are arguing over a case. One social worker argued vehemently that a child must come into care, which the team manager agreed with and then the other social worker argued that the child should remain with the family. The team manager agreed with this social worker too.

After the two social workers have left the room, the senior practitioner, felt compelled to say something, turned to the team manager and said: 'you have just given two diametrically opposite points of view – they can't both be right.'

The team manager nodded her head wisely and said: 'Yes and you are right too.'

Introduction

On the one side language and how people communicate is central to social work, as it is a profession which has had communication and interpersonal skills at its heart. Some of the 'classic' text books of social work practice e.g. Hargie (1997), Lishman (1994) and Trevithick (2000) placed communication skills at the heart of good practice. As a result social work is a profession which has a special interest in how language is used and what is said by practitioners to people both within and outside social work.

On the other hand words have been used to persecute and oppress, as explained by the superiority theory of humour and words such as *cripple* and *idiot* (just to cite two examples) are no longer in public usage (Philpot, 1999). Social work is a profession which has often been concerned with challenging the oppressive use of language and given both the interest in language and discrimination, it is not surprising that 'social workers have a reputation for being the most PC people around (although the term is so reviled that most disown the expression)' (Mickel, 2008).

Sasha Baron Cohen in an interview on 18th May 2012 on Radio 4 stated that 'from a history of persecution you must develop a sense of humour.' Freud collected jokes about Jewish people and Kulick (2000) collected gay jokes, and it could be suggested that the process of analysing or sharing a joke within one's own group in effect sanitises it. In a similar way Meachin (2013) suggested that the cartoon and radio series *Clare in the Community* offered a 'breezy riposte to the oft depressing media headlines about social work . . . lampooning social work's reputation for political correctness' (Meachin, 2013). Political correctness has often been a vehicle for humour and satire, and some examples are both senseless and amusing in themselves, although many of them quickly became urban myths repeated often by the right wing press, for example the allegation that *Christmas* and *New Year* was replaced with the term *Winterval* by Birmingham council (Philpot, 1999).

So how did the term political correctness develop and what does it mean in terms of humour and the practice of social work?

A history of political correctness

There are no clear origins of the term political correctness in British culture, although Mahadev Apte indicated that based on a study of the archives of the BBC that political correctness has been central to the censorship of British humour, since at least the 1930s and 'the BBC's files are full of admonitions to producers not to use jokes about asthmatics, crooked lawyers, "dagoes," "Confucius he say," spics, the Maltese, "niggers," effeminacy in men, stutterers, etc' (MacHale et al., 1997, p. 495). Most commentators pinpoint the late 1980s as the point at which the ironic, in-group connotations of the term 'PC' as used on the left were transformed into a blanket derogatory term with which to denigrate a plethora of left-liberal concerns (Johnson et al., 2003). This is supported by Weigle, who found that if you 'search databases of magazines and newspapers, you find that the phrase "politically correct" rarely appeared before 1990' (Weigel, 2016). Prior to this the phrase was used nearly exclusively within the political left, and almost always sardonically (Weigel, 2016). Political correctness, or at least the crime of political correctness has often been utilised by right wing and extreme right wing politicians to achieve their ambitions of attacking their political foes, most notably in recent years in America by Donald Trump, but also by Marine Le Pen of the Front National in France (Weigel, 2016). A trope popular on the British right holds that 'you can't talk about immigration these days thanks to PC taboos, which would be troubling except that British right-wing newspapers talk about it incessantly' (Burkeman, 2014).

Some researchers suggest that it is not unusual that it is predominantly male commentators who use political correctness as a vehicle to attack, as there is an inherent gender quality to oppressive humour use. Kazarian and Martin (2006) found that men enjoy and use humour to a greater extent than women and Watts (2007) argued that 'having a laugh' and being able to 'take a joke' is central to male identity, and that women appear to be much less comfortable with a 'joke culture' in the workplace. Women in Watts' (2007) research experienced joke telling as difficult to handle and she found that women's disdain for the excessive or inappropriate use of humour by male colleagues focused on its hostile nature and the damaging effects of jokes on other people. Not only is there a gender bias in humour use, but Longo (2010) has argued that inherently humour theory has a male bias and as a result is narrow in its focus.

Political correctness and social work

By the 1990s social work training in particular had come under attack from the press and others for its over-zealous approach to the teaching of racism and anti-racist practice. Tony Hall the then Director of the Central Council for Education and Training in Social Work was provoked to describe as 'nonsense' Bryan Appleyard's claims that the shortcomings of social work training and every-thing from bad social work practice in adoption cases to 'a wave of oppression and corruption' in British universities could be blamed on political correctness (Hall, 1993).

The criticism that social work is 'paralysed' by political correctness has not gone away and is often repeated in the wake of scandals e.g. Tim Loughton, the children's minister, who was criticising social work failings in the Bamu case of child abuse linked to 'witchcraft' 'said that a "wall of silence" was obscuring the full scale of cruelty in some communities where beliefs in evil spirits was com-mon . . . there has been only very gradual progress in understanding the issues over the last few years – either because community leaders have been reluctant to challenge beliefs which risk leading to real abuse in their midst; or because authorities misunderstand the causes or are cowed by political correctness' (Bing-ham, 2012). However in response Detective Superintendent Terry Sharpe of the Metropolitan Police, who led the Bamu case, said the central obstruction was a lack of awareness about witchcraft practices (Bingham, 2012).

After the Rotherham child sexual exploitation scandal Brewer (2014) com-mented that

> As worrying as the excess of political correctness in Rotherham, is the con-tinuing resistance of social work to controlled, sceptical, independent studies of its effectiveness. In contrast to medicine, that resistance has long been a very prominent feature of social work in Britain (and in most other coun-tries). Often, social workers not only resisted evaluation but blamed the mes-senger when the impact of social work interventions was shown to be either marginal or negative.
>
> (Brewer, 2014)

As such the overriding message from the media and politicians regarding social work is that it is dominated by 'political correctness' and poor practice which is caused by a disproportionate focus on issues of class, 'race' and gender (Lavalette and Penketh, 2013).

Social work is a profession with an explicit and established value base, which actively champions challenging discrimination (Banks, 2006; Thompson, 2006; Adams, 1996; Parrott, 2006). Many people are attracted to the practice of social work, including myself, because it has promoted a set of values which challenge discrimination and oppression. From its earliest days the political nature of prac-tice and a political consciousness are deep seated in social work and continues to be, as a self-proclaimed 'value based profession' which strives to 'promote social change and development, social cohesion, and the empowerment and liberation of people and principles of social justice, human rights, collective responsibility and respect for diversities are central to social work' (International Federation of Social Workers, 2014). Social workers have been most successful in achieving these goals when they have used evidence to show the extent and causes of social problems such as poverty and have challenged the institutions and the politicians who do not want such issues to be highlighted (Reisch and Jani, 2012), so we should not be surprised that social work continues to be criticised for its political nature and attacked by the press and politicians.

Reisch and Jani (2012) also argue that practitioners have become 'frightened by media attacks on political correctness' (Reisch and Jani, 2012, p. 1141) and Polly Neate writing in Philpot's edited book on social work and political correctness 'social workers themselves have put up no kind of fight to defend their ideals; they have not even made much of an attempt to explain them' (Neate in Philpot, 1999, p. 65). Political correctness is a term which undoubtedly has been used to restrict social work: 'political correctness can be a term by which you beat people over the head with . . . you think "well excuse me what exactly is wrong with saying that," in the liberal context, people should have an open and accepting attitude towards others in the community. What exactly is wrong with that?' (Orford, 2018).

This leaves practitioners with some dilemmas and in the next section I highlight some of the dilemmas which impact on practice.

Practice issues and political correctness

Many comedy shows have been criticised for excluding women and there are stereotypes around women not being good at comedy, but women in social work do 'do humour' and particularly anecdotal humour e.g. the King and Brown's collection of stories, *Brummie girls do social work* (2015). Given the gendered employment division of social work, i.e. 75% female workforce (Galley and Parrish, 2014) often working with predominantly female service users, I was once told a joke by a female colleague aimed specifically at men working in social work (which challenges the gender stereotype):

Q: Why do male social workers prefer briefs to boxers?
A: Their 'boys' prefer a warm, supportive environment!

Most social workers are women (Galley and Parrish, 2014) and feminism has played an important role in the development not only of social work and its value base, but also some authors argue in the 'discovery of child abuse and domestic violence.' For some practitioners challenging sexism is an ongoing contest and one practitioner commented:

> *One of his colleagues* [thought it was funny as he] *had been disciplined because he insisted on calling his team members 'ladies.' . . . and I thought*

'what is funny, I don't want to be called the lady.' . . . I explained that it was not anything to do with being lesbian but a male caricature of what it is to be a woman. I think characterising 'ladylike' behaviour is a male definition of pleasing somebody else. We had a discussion about it and of course I got an e-mail immediately afterwards which began with 'ladies' as an address. I did not respond to the e-mail except to say I really resent this. They probably thought it was a humour thing misplaced, but I am the only social worker there.

(Jordan, 2015)

One could suggest that these comments fit the trope of the politically correct, but socially uptight and humourless practitioner, a stereotype which the *Clare in the Community* series exploits. However, she made a serious and important point that gender oppression is constructed by and dependent upon language and as such can be challenged through language. The HCPC Standards of Proficiency (2017) require social workers to be able to practice in a non-discriminatory manner and to be able to use practice to challenge and address the impact of discrimination, disadvantage and oppression, particularly when language reinforces cultural stereotypes. It could be argued that the worker is practicing appropriately by challenging sexist language. As a lone social worker amongst other professionals the comment also revealed the workers feelings of isolation and vulnerability, particularly when she tried to challenge her male colleagues. This indicates that challenging inappropriate and offensive humour can be an isolating experience, for the individual.

Similarly another worker commented on the political culture and context of the office in which he had practiced and where this became uncomfortable:

When I was working in [local authority] *it was a very politically sensitive office. You just could not make . . . jokes. That became a difficult place to work and there was lots of intercultural dynamics and fractious groups . . . I found that a very difficult place to work as I could not make the remarks that I wanted to . . . I thought there was a sense of fear there, there was times bullying amongst the groups that just made people retreat . . . people retreated and would not make statements for fear of being singled out.*

(Jordan, 2015)

Such comments appear to reinforce the media portrayal of social work offices as controlled and harsh environments, full of fearful workers, no longer able to make jokes for fear of the consequences. Both practice examples support the notion that engaging in humour can be socially isolating, just as is the process of challenging inappropriate humour. Overall this shows that any humour use involves an element of risk taking and that some offices can be safe places in which to risk humour. Some social workers suggested that practitioners needed to make statements which were considered unPC, or not 'politically correct:'

If we said something that was perhaps a little bit unPC, it was only because we . . . were really struggling to cope with it . . . and in the safety of the office, that is how we coped.

(Jordan, 2015)

This is one aspect of the paradox of humour, that it is necessary for working relationships, but it can also be used to alienate and oppress. It seems to me that we need a language to communicate in these ways or to find safe places. I suspect as Orford found, social workers in the end had a realistic approach to this:

I can remember decades ago a training session run by an amazing, really, really excellent trainer, a black bloke, and he stood up in front of us and he said 'I was walking down the street and this woman dropped her stuff and I went over to help her out and she said "oh, thank you very much we don't get many niggers around here"'and he said 'the response should be "oh I should be offended" but I am talking to an 80-year-old woman, who clearly didn't mean to be offensive in any way . . . she was using language that was no longer appropriate, but she was an elderly woman who had been brought up in a different time and she wasn't meaning to be offensive.'

(Orford, 2018)

This comment reinforces the point that humour is located within a wider social and cultural framework. Social work has a long history of challenging oppression (Banks, 2006; Thompson, 2006) which is central to the profession's value base. The key point that frames social work behaviour in relation to humour is that it must be within a discourse which is sensitive and accountable. It could be suggested that the contrasting comments above reveal that there are two processes operating: first that social work is insecure about its position in regard to language and PC accusations, and second that paradoxically social work is both *a part of and apart from* society, in the sense that everyone understands and uses humour, but that social workers are not always sure of how to apply and where to use it.

Conclusion

Social work has for most of its history been on the side of the oppressed and marginalised, and the profession should be proud of its history of challenging discriminatory language and oppressive practices. It is unsurprising that social work should be attacked as 'paralysed' or 'humourless' because the reality is that social work is not going to be popular with some politicians and some of the media.

So is political correctness a phantom enemy for the right, as Weigel (2016) suggests? At times concerns about the impact of practice on marginalised groups may have 'paralysed thinking,' but social work has correctly been concerned with doing the least harmful thing and being sensitive to language. Being concerned over whether something said or done has given offence is proper and correct. The reality is that language changes all the time and this is part of an historical process and a changing political landscape. The profession has been part of that progressive force for good, and it is inevitable that social work will be on the side of being criticised for being politically correct. So rather than disowning the phrase politically correct as Mickel (2008) suggested, social work should instead stand up and be proud of this history and cherish the campaigns for fairness and equality, which have been founded in political correctness.

References

Adams, R. (1996) *Social Work & Empowerment* Basingstoke: Palgrave Macmillan.

Banks, S. (2006) *Ethics and Values in Social Work* Basingstoke: Palgrave Macmillan.

Bingham, J. (2012) Witchcraft child abuse: Social services and police 'cowed by political correctness' claims minister *The Telegraph* www.telegraph.co.uk/news/religion/9475115/Witchcraft-child-abuse-social-services-and-police-cowed-by-political-correctness-claims-minister.html

Brewer, C. (2014) Rotherham has proved it again: Social work just doesn't work: This profession resistant to empirical evaluation may harm as much as it helps *The Spectator* August 30 www.spectator.co.uk/2014/08/the-rotherham-report-suggests-that-social-workers-are-as-often-harmful-as-helpful/ (accessed 25/7/18).

Burkeman, O. (2014) Political correctness really works! Sorry, conservatives, but science just said so *The Guardian* Thursday November 13 www.theguardian.com/commentisfree/oliver-burkeman-column/2014/nov/13/political-correctness-science-conservatives-liberals (accessed 27/7/18).

Galley, D. and Parrish, M. (2014) Why are there so few male social workers? New research shows more male students are dropping out and that they may be singled out by lecturers as a rare breed *The Guardian* www.theguardian.com/social-care-network/2014/jul/25/why-so-few-male-social-workers (accessed 27/7/18).

Hall, T. (1993) Letter: Social work and political correctness *The Independent* Thursday August 5 www.independent.co.uk/voices/letter-social-work-and-political-correctness-1459459.html (accessed 25/7/18).

Hargie, O. (Ed.) (1997) *The Handbook of Communication Skills* (2nd ed.) London: Routledge.

HCPC (2017) Standards of proficiency: Social workers in England London: Health Care Professions Council.

International Federation of Social Workers (2014) *Global Definition of Social Work* www.ifsw.org/what-is-social-work/global-definition-of-social-work/ (accessed 27/7/18).

Johnson, S., Culpeper, J. and Suhr, S. (2003) From 'politically correct councillors' to 'Blairite nonsense': Discourses of political correctness in three British papers *Discourse and Society* 14(1), 29–47 [0957–9265 (200301) 14(1), 29–47; 028928].

Jordan, S. (2015) *That joke isn't funny anymore: Humour, jokes and their relationship to social work.* London: University of East London. Professional doctorate thesis.

Kazarian, S. S. and Martin, R. A. (2006) Humor styles, culture related personality, well-being, and family adjustment among Armenians in Lebanon *Humor: International Journal of Humor Research* 19(4), 405–423.

King, S. and Brown, N. (2015) *Brummie Girls Do Social Work* Oxford: Shrewsbury You-Caxton Publishers.

Kulick, D. (2000) Gay and lesbian language *Annual Review of Anthropology* 29, 243–285.

Lavalette, M. and Penketh, L. (Eds.) (2013) *Race, Racism and Social Work* Bristol: Policy Press.

Lishman, J. (1994) *Communication in Social Work* Basingstoke: Palgrave Macmillan.

Longo, M. (2010) Humour use and knowledge-making at the margins: Serious lessons for social work practice *Canadian Social Work Review/Revue canadienne de service social* 27(1), 113–126.

MacHale, D., Nilsen, A. P., Derks, P., Lewis, P., Berger, A., Mintz, L., Nilsen, D. L. F., Gruner, C., Oring, E., Ruch, W., Morreall, J., Attardo, S. and Ziv, A. (1997) Humor and political correctness *Humor* 10(4), 453–513. https://doi.org/10.1515/humr.1997.10.4.453

Meachin, H. (2013) Media watch: How clare in the community offers a welcome chance for a laugh *Professional Social Work* February.

Mickel, A. (2008) Political Correctness and Its Effect on Practice *Community Care* www.communitycare.co.uk/2008/12/01/political-correctness-and-its-effect-on-practice/

Orford, F. (2018) Interview with the author 12th April 2018.

Parrott, L. (2006) *Values and Ethics in Social Work Practice* Exeter: Learning Matters.

Philpot, T. (Ed.) (1999) *Political Correctness and Social Work* London: 1EA Health and Welfare Unit, p. 82. ISBN 0 255 36457.

Reisch, M. and Jani, J. S. (2012) The new politics of social work practice: Understanding context to promote change *British Journal of Social Work* 42, 1132–1150. DOI: 10.1093/bjsw/bcs072

Thompson, N. (2006) *Anti-Discriminatory Practice* Basingstoke: Palgrave Macmillan.

Trevithick, P. (2000) *Social Work Skills: A Practice Handbook* Buckingham: Open University Press.

Watts, J. (2007) Can't take a joke? Humour as resistance, refuge and exclusion in a highly gendered workplace *Feminism & Psychology* 17, 259.

Weigel, M. (2016) *Political Correctness: How the Right Invented a Phantom Enemy* London: Guardian www.theguardian.com/us-news/2016/nov/30/political-correctness-how-the-right-invented-phantom-enemy-donald-trump (accessed 10/4/18).

7   Rottweilers and subversives
Problematic and anti-social humour

Q: How many social workers does it take to change a light bulb?
A: The light bulb doesn't need changing; it's the system that needs to change.

Introduction

This chapter provides examples of hostile jokes and humour, as social work often finds itself the target of jokes and humour. Humour and challenges to authority go hand in hand, e.g. in an interview in 2017 Nish Kumar stated that 'The root of laughter has got to be sticking it to someone. Even when you're a kid, the first time that something makes you laugh is when the authority of your parents is undermined, or your teachers' (Brinkhurst-Cuff, 2017). Jokes and humour can be powerful tools in creating change, as they can be linked to resistance and to challenging oppression. Joking relationships are not presumed to be necessary for the continuation of social life in general, but humour can ease the exercise of power, as people who are adept at using humour can adopt the role of 'sage fool' as a way of managing and expressing dissenting opinions (Cooper, 2008), so humour could be a means by which 'subordinates' challenge power structures and make what might be thought of as 'risky statements.' Anthropological studies of oppressed groups' use of humour suggest that jokes, joking behaviour and humour flourish under repression, as people seek ways to express themselves through humour and to undermine their oppressors.

It is not possible, I would argue, to consider humour in all its myriad of expressions without too encountering the times when humour has been used in hostility to social work. The danger is that humour can be offensive and problematic for that reason. Social work is a profession committed, at least by its ethical standards and ascribed value base, to not being offensive, indeed it is one of the core requirements, but given that so much humour can be characterised as offensive what happens in social work? This chapter examines the time when humour is problematic and considers the value of what could be termed subversive humour.

Social 'subversiveness' and humour

Social work has a long history of political subversion and challenging existing power structures, particularly where they oppress minorities and the most vulnerable in society. Lavalette (2011) and others have termed this 'speaking truth to power' (Lavalette, 2011, p. 6). Humour too has the potential to subvert forms of dominance and asserting rebellious or counter-managerial views in the workplace.

For some social work is a hotbed of subversiveness and 'marxist insurgents urging service users, students and fellow practitioners to over throw capitalism' (Phibbs, 2012). The commentator Harry Phibbs felt that social work courses indoctrinated their students using 'degrading thought control' and 'endless stuff about...how the capitalist system must be overthrown' (Phibbs, 2012). However at its roots the early pioneers of social work practice such as Octavia Hill and Jane Adams were challenging the status quo and could equally be charged with being subversive in their approach to the excesses of capitalism. This suggests that the pioneers of social work were as subversive and radical in their approach and that from its inception social work was about challenging unfairness and inequality. BASW, viewed in the past as the antipathy to radical social work (Lymbery, 2001), has more recently found itself on the side of subversion. For example Ruth Cartwright, BASW England's Manager, argued that 'If you believe social work should be a force for change, then it should be exposing rather than concealing suffering and lack of humanity . . . it should be speaking with and for service users and those who are discriminated against' (Robb, 2012).

This does not sit comfortably with the government and ministers, as a 'subversive' is referred to as a traitor by the government in power. Subversion comes in many forms and in the 2000s the government was concerned with the subversive activities of environmental activists to the point that the police under the National Public Order Intelligence Unit (NPOIU) infiltrated several groups and undercover police officers had children with activists (Bingham, 2011).

Political subversion proliferates under conditions of oppression and repression. Subversion is often a goal of professional comedians, who seek to question the established order. In this respect satire as a form of humour is one of the most common forms of subversion. For example Jasper Carrott told stupidity jokes about the readers of Sun newspaper in an attempt to subvert the power and influence of the newspaper, but the jokes never caught on (Davies, 1998). The political joke has been seen as the most widespread form of subversive humour is which, in the words of Benton (1988) takes place in 'modern dictatorships of all political sorts' (Benton, 1988, p. 33) and Benton suggests, it is almost obvious that any dictatorship leads to a considerable production of not only jokes but various forms of humour. There are plenty of examples of subversive political humour. The economic and political crisis in Zimbabwe associated with Robert Mugabe's presidency led to a flourishing of internet humour and Musangi (2012) found that 'humour can be employed as a form of subversion especially in contexts of autocracy' (Musangi, 2012, p. 162). Orwell said that 'every joke is a tiny revolution' (Lewis, 2009, p. 19) and Lewis (2009) argued that jokes are powerful tools in creating change, as they are linked to resistance and to challenging oppression. Jokes and humour aimed at making fun of the oppressively powerful appear 'justified' and the jokes, which punctured the pomposity of the Soviet Union, arguably worked to eventually undermine the social authority of the regime (Lewis, 2009).

It is unlikely that any government will look favourably on a profession whose value base and ethical position is associated with challenging poverty, inequality and oppressive practices, particularly if the policies of the government are those causing the poverty and inequality in the first place. This places the profession of social work often at odds with the government of the day and particularly the recent administrations, whose austerity policies have had such significant negative effects on service users and public services as a whole. It is not unusual therefore that social work has been undermined by successive governments, who have sidelined social work, subsuming it under 'social care,' and removed its central role in relation to both adult and youth offending and in mental health (Jones, 2012; Rogowski, 2012; Warner, 2013). Warner (2013) has shown how politicians, in conjunction with the press, actively mobilised public anger towards social work, to meet political ends, often to sideline and reduce social work's role and authority in society.

People who use humour to challenge are as Bergson said 'disguised moralists' (cited in Coser, 1959) and Holmes (2000) found that humour could be a means by which subordinates could challenge power structures and make what might be thought of as 'risky statements,' in a lighthearted way. Christie (1994) argued that humour enables us to tolerate antithetical ideas and 'Plato saw how comedy and laughter could undermine the rulers of his Republic' (Ritchie, 2010, p. 161). Commentators have concluded that social work's capacity to challenge power structures and 'be a force for progressive policy and social change had been significantly eroded' (Naqvi, 2013), and given this erosion of social work's capacity to subvert and challenge the government's policies and the changes it has been making to the welfare state, what happens to social workers who express subversive ideas?

In terms of humour made at the government, it's more common to find such examples in the work of the professional humourists, one example being Harry Venning with *Clare in the Community*. As indicated in Chapter 4 *Clare's* targets are frequently the Tory government and their continued cuts to services. The punchlines in *Clare* are often ironically subversive. An example of this can be found in the cartoon published in the *Guardian* on 20th November 2012, in which a mother, who is feeding a screaming child in an impoverished and messy bedsit, says to Clare 'It's when the Olympics "feel good factor" finally runs out that worries me.' Further examples of the subversive political humour can be found in the TV series *Damned*. In an interview published on-line Jo Brand spoke explicitly about the Government cuts which were hitting social work 'The demands on the workers are ever magnifying, and the time available to do them therefore is shrinking. . . . And it's not just direct cuts to social services, its things like Sure Start children's centres. And loads of those have closed down, so that whole rung of support has disappeared. So mothers who are trying to raise their kids on virtually no money and with no support, don't have those places to go to that they used to' (Brand, 2018). In this respect the series has a subversive 'message underneath the surface' (Brand, 2018).

Social workers use subversive humour to challenge the authority of the managers. In this context humour in social work is similar to how many workers in other industries use it to subvert the power structures and 'undermine management control . . . [and] can be characterized as the 'humour as worker resistance" (Butler, 2015, p. 43). Managers, as the figures of authority in the offices, were often targets for subversive humour and this provides some examples of the relationship between subversion, humour and social work. In response to an article by Drinkwater (2011) about the importance of humour in social work offices, this comment was made:

> Sadly I jest not when I say half plus of my team are on antianxiety/ antidepressants as a result of the toxic adult care team we are in. On the funny side senior management are a joke and provide the odd smile. One of the managers is the spitting image of David Dickinson's love child with a large dollop of Elvis impersonator, the other looks and acts like a used car salesman.
>
> (Community Care Space, 2011)

Whilst this comment has practice issues too with reference to the 'toxic' and challenging environment of a social work team, this was reinforced by other comments which highlighted oppressive management styles:

> Our office is suppressed by a middle management level who sit at the top of the office, monitoring who is doing what and saying what to whom :(This is a shame because when this level of middle management is not in the office, we get more work done than when they are there, humour at these times is rife, and their time of absence from the office is used by staff to offload the stresses and frustrations of the job. Essentially: Laughing is banned in our office between the hours of 9–5, and it's also the case that if anything is said which is funny or irrelevant to the workplace, it is noted, and brought up in supervision; along with a copy of the acceptable behaviour policy.
>
> (Jordan, 2015)

The concerning aspect of this is that with this level of surveillance, social workers are unable to 'do' humour in front of middle management and can only offload their stress and frustrations of the job in the absence of middle management. It is possible that this is an office where the need to express oneself are not possible, which lacks the necessary emotional support, or emotional containment. Emotions then have to be expressed in secret when the manager and authority figures are not around. It suggests that some social work offices are stuck in rather out-dated management practice and illustrates Karlsen and Villadsen's (2015) point that 'in traditional management discourse, humour in the workplace is often viewed as undermining productivity and subverting the maintenance of authority' (Karlsen and Villadsen, 2015, p. 517). So whilst social workers used humour to subvert authoritarian management, service users turned to humour too to express their own subversive feelings about social work.

Humour, hostility and fear

Social workers are likely to be working with many people who are struggling to cope, and given that 'laughter is the flip side of fear' (Smith, 2013), individuals who are fearful about their experiences of the care system have published hostile jokes about social workers on-line in what could be suggested are attempts to subvert the profession's authority. Scott (1985) described joking as a 'weapon of the weak.' Facebook and other online technology have allowed parents to voice their unhappiness with social work practice through jokes:

> *A woman stood and watched a social worker being beaten by ten people, after a policeman broke them apart he said to the woman, 'why didn't you try to help'? To which she replied, 'I thought ten was enough.'*
>
> Source: nojusticeforparents webpage
> http://staffordshiresocialservices.wordpress.com

This joke has been used oppressively for some time e.g. in the 1970s it was often used by Les Dawson with his 'mother-in-law' standing in for the social worker. People look towards humour as a relief, but it's often uncomfortable and problematic, and hostile subversive jokes, voiced by parents who feel aggrieved, are painful reminders of the fear and hatred social workers experience. They can be subversive to social work power and no less comfortable for that. There are several jokes that fall into the hostile/ fearful theme, and as in the example above making humour out of jokes about physical attacks on social workers. For practitioners it is both painful and fearful to read such jokes, and the jokes reflect a particular view of social work practice. Here the humour is used as a physical attack on the profession of social work, a way of expressing the hatred and fear parents or people who have no power use to challenge authority, most notably in child protection work. The aggressive dog such as the Rottweiler or Pit Bull Terrier are a familiar feature in the lives of many children and families and are used as a status symbol (Donovan et al., 2013). In this context this infamous joke appeared in the late 1980s:

Q: What's the difference between a social worker and a Rottweiler?
A: You have a chance of getting your child back from a Rottweiler

It could be suggested the joke was driven in part by increasing public and political concern over dangerous dogs and particularly attacks on children by violent dogs which were kept as a status symbol by some families. The original joke arose from this context, and then underwent several transformations and a more aggressive version of the joke appeared in the 1990s:

Q: What is brown and black and looks good on a social worker?
A: A Rottweiler

These jokes took on a particular poignancy in the case of Peter Connelly, when Jason Owen had moved into Tracey Connelly's house with his 15-year-old girlfriend and five children, aged variously from seven to 14, together with a pet snake, and at a later date he also added a Rottweiler dog to the household (McShane, 2009). By the late 2000s the dog in the joke had metamorphosed from a Rottweiler into a pit bull terrier, which reflected the changing nature of dog ownership in Britain. These jokes fall within the superiority theory of humour, where the hostility conveyed by the humour involves mockery and derision, but also, I suggest represents a particular form of the superiority theory, that is subversion. Such jokes reflect explicit anger towards the profession, albeit for a minority of people who have negative experiences of social work. A final example of this type of joke suggested a poignant reflection on the 'unhappy' lives of social workers:

Q: What's the difference between a social worker and a Rottweiler?
A: These days a Rottweiler has more chance of being rescued from a life of misery?

As a practitioner I had sometimes heard anecdotal reports of the RSPCA removing pets from families, where they were deemed to be hurt or neglected, but then children were left in the care of those same families. It is possible to conclude that the hostility and aggression revealed by the jokes encapsulate some of society's frustrations and hostility to the actions of the profession. Threats, hostility and physical attacks on social workers are real and common (McGregor, 2010; Donovan et al., 2013). Practitioners' fear of violent men remains a significant theme in many inquiry reports and serious case reviews, in the sense that fear of the violent abuser can paralyse thinking and stop practitioners acting to save children (Brandon et al., 2008). Stanley and Goddard (2002) found that feelings comparable to helplessness were a daily feature of many child protection practitioners' lives, and that children were being left in dangerous situations. As a result it should not be surprising that social workers have been accused of losing their sense of humour.

'Losing their sense of humour'

The internet now is a vehicle for some people to express their thoughts of child protection services and an example of this is a British website entitled Child Protection Resource run by a Sarah Phillimore, a family lawyer. The site indicates its aim is to 'help everyone who is involved in the child protection system, in whatever capacity by providing up to date information about relevant law and practice, and contributing to the wider debate about the child protection system' (Phillimore, 2017). On the site a question was posed on 7th April 2017: 'Why don't social workers have a sense of humour?' There then followed an example from a birth parent

> We were asked some intrusive questions about our sex lives and we tried to make a joke about it. It would have really helped if the social worker could have reacted in a more relaxed way, rather than making it obvious that she

was shocked and upset by what we said. It goes beyond 'having a sense of humour' I really noticed that everything we said or did was seen in the most negative light possible. So making lighthearted comments or jokes was used against us. I know this is a serious situation and it isn't always the right thing to try and joke about. But sometimes if we were scared or nervous we would try and lighten the mood. But anything we said that we thought was obviously a joke was taken seriously. My partner jokingly kissed my neck and scooped me into his arms during an assessment. The assessor wrote that she thought we were intending to have sex in the office! and that we probably indulged in 'inappropriate sexual activity' in front of our child.

(Phillimore, 2017)

As a practitioner it's impossible not to feel a sense of injustice on behalf of the parent who gave this example. In response to the example its worth stating a defence about the need for humour and even in the face of subversive attempts for practitioners to maintain their humanity and recognise the unifying power of humour and in the words of one respondent to the website: 'it would be odd if professionals trained in human psychology, to some degree, didn't recognise the function of humour' (Phillimore, 2017).

It's important to remember that 'humour is, indeed, a powerful transmitter of the popular mood in societies where this mood can find no officially sanctioned outlet' (Benton, 1988, p. 33) and it could be argued that hostile jokes and com-ments cited in this section communicate something about some service users' concerns about social work practice. The work of Parr and Nixon (2009) suggest that this use of humour represents frustration and hostility which might be linked to the public discourse which advocated tough sanctions on anti-social families. As such the social work profession is a target for some families' anger as social workers were effectively co-opted into challenging families whose response is to engage in anti-social work humour as one of the last forms of outlet left to them.

Social workers' problematic use of humour

Studies of social workers in practice can reveal examples of where social workers have used humour oppressively. Mik-Meyer (2007) in her study of social workers in a Danish rehabilitation centre found that social workers often shared jokes and made humorous remarks about clients, but these were often ignored or met with silence by the clients they were working with. She found that that social work-ers' discussions of clients' personalities 'were humiliating' (Mik-Meyer, 2007, p. 13), and in the study Mik-Meyer found that the social workers used humour to give them power over their clients, and the humour use simply 'reinforced the difference between social workers and their clients' (Mik-Meyer, 2007, p. 16). On other occasions social workers used irony and sarcasm because they had a 'hard time helping some clients who might be lying' or ill-suited to the labour markets (Mik-Meyer, 2007, p. 19), but at the same time such behaviour was accompanied by embarrassment on the part of the social workers, and when challenged this was justified by the social workers as 'ridding themselves of the emotional shit the clients bombard us with all the time' (Mik-Meyer, 2007, p. 19).

In another example five social workers in Scotland and England were dismissed or admonished by the General Social Care Council in 2010 after they forwarded a string of 'joke' e-mails, including one containing a mocked-up image of convicted sex offender Gary Glitter carrying a child in a plastic bag. Forwarding of 'joke e-mails' is a common experience in most modern workplaces, and homes, where the use of the Internet is commonplace. In a study from 2005 about a third of internet users in the UK used the internet to find jokes, cartoons and other humorous material (Dutton et al., 2005), but for social workers the consequences of telling an offensive or oppressive joke has significant consequences and it's appropriate that such behaviour should be challenged and sanctioned.

In discussing the forwarding of joke e-mails in social work offices, respondents to the CareSpace forum suggested that many social work practitioners wish to be taken seriously and are earnest in their endeavours to help and support service users, and for many practitioners the forwarding of joke e-mails is unacceptable e.g. 'What they did was totally inappropriate . . . The fact it's an agency that seeks to protect children makes it even worse.' The responses suggested that forwarding joke e-mails in social work offices is contingent on them (a) providing some humorous relief and most importantly (b) not making exploitative or oppressive comments.

Practice issues

So some social workers engaged in inappropriate humour use, subverting their code of ethics and they were quite properly sanctioned.

In the context of an oppressive management style the lack of humour at work could be equated with higher experiences of stress and levels of absenteeism amongst staff:

> *Have worked in offices where management frowned on any 'irrelevant' chat never mind humour – result = high absence rate, stressed out staff and high staff turnover. Worked where we were treated like adults who were capable of judging what was appropriate, what (even if dark humour) was needed to lift the mood, and when it was Ok to have a coffee and a chat – result = high retention of staff, low sick rate, and guess what – the work got done, and sometimes we even went the extra mile and were happy to do so! Would be bosses – take note!*
>
> (Jordan, 2015)

Online comments often criticise the way social work is managed, and some of the comments posted to the Community Care Forum discussion board suggested that high levels of scrutiny forced humour below the surface (or at least after the managers have left the office). There is a dialectical process revealed: on the one hand humour is tolerated and longed for by social workers themselves, as a form of emotional release, but on the other management appear at times to suppress any subversive, rebellious and 'inappropriate' outbursts of humour, which appear to seep out when 'management is not looking.'

As Carpenter (2011) has suggested this scenario will be familiar to many social work practitioners. In such offices misery, suspicion of management and pervasive negative mindsets can become the dominant mood and behaviour of the teams. Such environments can take their toll on even the most positive, buoyant and resilient social workers, and may go some way in explaining the high turnover and low retention rates in the profession.

Conclusion

Subversive humour is complex and paradoxical, as it serves several social purposes (often at the same time) as it is both social and anti-social – it can bring people together, but at the same time through mockery can exclude people and Billig (2005) found that humour 'can function to protect the social order, keeping social actors in line, but simultaneously it can express pleasure at subverting that same order' (Billig, 2005, p. 235).

Jokes and humour are sometimes both the 'resistance and the attack' which reflects the ambivalence in which social work is held. It is troubling that social work finds itself in both a unique and a complex role to carry out, and the subversive aspect of humour leads to the conclusion that humour is a mechanism through which social workers negotiate their place in the world. The ambivalent and complex role that humour plays in the world reflects the ambivalent and troubled place that social work too occupies. Social workers fear not being taken seriously, but conversely use subversive humour and jokes to manage their work, and sometimes use humour inappropriately and oppressively.

Humour's 'special powers' rest in the fact that it has both intellectual and emotional sides (Korte and Lechner, 2013), and it's possible that subversive humour has a unique role in enabling practitioners to challenge oppressive practices, at a time when 'there is on the one hand areas of closing down of the space for worker discretion through increased managerialism and modernisation, and both citizens and workers react against restriction and regulation in ways that subvert government intentions' (Barnes and Prior, 2009, p. 10) and one such reaction is the use of humour both by social workers and citizens.

References

Barnes, M. and Prior, D. (2009) *Subversive Citizens: Power, Agency and Resistance in Public Services* Bristol: University of Bristol, Policy Press.

Barron, J. W. (Ed) (1999) *Humour and Psyche: Psychoanalytic Perspectives* Hillsdale NJ: The Analytic Press.

Benton, G. (1988) The origins of the political joke in Powell, C. and Paton, G. E. C. (Eds.), *Humour in Society: Resistance and Control* London: Palgrave Macmillan.

Billig, M. (2005) *Laughter and Ridicule towards a Social Critique of Humour* London: Sage Publications.

Bingham, J. (2011) Mark Kennedy: 15 other undercover police infiltrated green movement *The Telegraph* www.telegraph.co.uk/news/earth/earthnews/8262746/Mark-Kennedy-15-other-undercover-police-infiltrated-green-movement.html

Brand, J. (2018) Social workers are not these weird, useless, out-of-touch characters . . . *Interview with Chortle* www.chortle.co.uk/interviews/2018/01/30/39017/social_workers_are_not_these_weird,_useless,_out-of-touch_characters... (accessed 31/7/18).

Brandon, M., Belderson, P., Warren, C., Howe, D., Gardner, R., Dodsworth, J. and Black, J., University of East Anglia (2008) *Analysing child deaths and serious injury through abuse and neglect: What can we learn? A biennial analysis of serious case reviews 2003–2005* Research Report, DCSF-RR023 London: Department for Education and Skills (DCSF).

Brinkhurst-Cuff, C. (2017) Let's talk: A conversation special: Nish Kumar meets Reni Eddo-Lodge: 'I've come to you for reasons to be cheerful: Go!' The comedian and the writer discuss race, humour and staying positive *The Guardian* December 2.

Butler, N. (2015) Joking aside: Theorizing laughter in organizations *Culture and Organization* 21(1), 42–58. DOI: 10.1080/14759551.2013.799163

Carpenter, J. (2011) *A Response to Drinkwater, M. (2011) On Reflection: Mark Drinkwater on Humour in Social Work* https://mypotentio.bloomfire.com/posts/6593-humour-in-social-work/public

Christie, G. L. (1994) Some psychoanalytic aspects of humour *The International Journal of Psychoanalysis* 75, 479–489.

Community Care (2011) *Care Space Forum* www.communitycare.co.uk/join-social-work-online-community/ (Forum closed in 2013).

Cooper, C. (2008) Elucidating the bonds of workplace humor: A relational process model *Human Relations* 61, 1087–1115.

Coser, R. L. (1959) Some social functions of laughter: A study of humour in a hospital *Human Relations* 12, 171–182.

Davies, C. (1998) *Jokes and Their Relation to Society* Berlin & New York: Mouton de Gruyter.

Donovan, S., Donovan, T., Smith, R., Brody, S., Pemberton, C., McGregor, K. and Samuel, M.(2013)*Death Threats and Dogs: Life on the Social Work Frontline* (Kindle ed.) Community Care.

Drinkwater, M. (2011) *On Reflection: Mark Drinkwater on humour in social work* http://www.communitycare.co.uk/2011/01/06/why-humour-is-so-important-in-the-social-work-workplace/ (accessed 06/01/2011).

Dutton, W. H., Di Gennero, C. and Hargrave, M. A. (2005) *The Oxford Internet Survey (OxIS) Report 2005: The Internet in Britain* Oxford Internet Institute, Oxford University.

Holmes, J. (2000) Politeness, power and provocation: How humour functions in the workplace *Discourse Studies* 2, 159–185.

Jones, R. (2012) The best of times, the worst of times: Social work and its moment *British Journal of Social Work* October 8. DOI: 10.1093/bjsw/bcs157

Jordan, S. (2015) *That joke isn't funny anymore . . .: An exploration of humour, jokes and their relationship to social work* Professional Doctorate in Social Work the Tavistock/Cass School of Education and Communities.

Karlsen, M. P. and Villadsen, K. (2015) Laughing for real? Humour, management power and subversion *Ephemera Theory and Politics in Organization* 15(3), 513–535 www.ephemerajournal.org

Korte, B. and Lechner, D. (Eds.) (2013) *History and Humour: British and American Perspectives* Bielefeld: Transcript Verlag.

Lavalette, M. (Ed.) (2011) *Radical Social Work Today: Social Work at the Crossroads* Bristol: University of Bristol, Policy Press.

Lewis, B. (2009) *Hammer and Tickle* London: Orion Books.

Lymbery, M. (2001) Social work at the crossroads *The British Journal of Social Work* 31(3), June 1, 369–384. https://doi.org/10.1093/bjsw/31.3.369

McGregor, K. (2010) Stabbed social worker had no warning of death threat *Community Care* Friday December 10.

McShane, J. (2009) *It Must Never Happen Again: The Lessons Learnt from the Short Life and Terrible Death of Baby P* London: John Blake Publishing.

Mik-Meyer, N. (2007) Interpersonal relations or jokes of social structure? *Laughter in Social Work Qualitative Social Work* 6(9).

Musangi, J. (2012) Chapter 8 'A Zimbabwean joke is no laughing matter': E-humour and versions of subversion in Chiumbu, S. and Musemwa, M. (Eds.), *Crisis! What Crisis? The Multiple Dimensions of the Zimbabwean Crisis* Cape Town: HSRC Press www.hsrcpress.ac.za

Naqvi, S. (2013) Can you be a Tory and a social worker? *Professional Social Work* March http://cdn.basw.co.uk/upload/basw_123916-1.pdf

Parr, S. and Nixon (2009) Chapter 7 Family intervention projects: Sites of subversion and resilience in Barnes, M. and Prior, D. (Eds.), *Subversive Citizens: Power, Agency and Resistance in Public Services* Bristol: University of Bristol, Policy Press.

Phibbs, H. (2012) Social work training is where the seeds of scandal are sown *Conservative Home* website published on line November 26 www.conservativehome.com/localgovernment/2012/11/social-work-training-is-where-the-seeds-of-scandal-are-sowed.html

Phillimore, S. (2017) *Why Don't Social Workers Have a Sense of Humour?* http://childprotectionresource.online/why-dont-social-workers-have-a-sense-of-humour/ (accessed 31/7/18).

Ritchie, C. (2010) Against comedy *Comedy Studies* 1(2), 159–168.

Robb, B. (2012) Prejudiced attack on social work training ignores real issues facing profession *Community Care* www.communitycare.co.uk/2012/11/30/prejudiced-attack-on-social-work-training-ignores-real-issues-facing-profession/

Rogowski, S. (2012) Social work with children and families: Challenges and possibilities in the neo-liberal world *British Journal of Social Work* 42(5), 921–940.

Scott, J. (1985) *Weapons of the Weak: Everyday Forms of Peasant Resistance* New Haven: Yale University Press.

Smith, H. (2013) From Greek tragedy springs a new generation of comedy *The Guardian* December 28.

Stanley, J. and Goddard, C. (2002) *In the Firing Line: Violence and Power in Child Protection Work* Chichester, West Sussex: Wiley.

Warner, J. (2013) Social work, class politics and risk in the moral panic over Baby P *Health, Risk & Society* [Online] 15, 217–233. http://dx.doi.org/10.1080/13698575.2013.776018

8 Conclusion

Choosing between our comic or tragic potentials

> SW to client: 'Anxious, depressed, unsure about the future . . . still enough of my problems, what can I do for you?'
>
> (thanks to Fran)

Humour is not one single thing

A key position of this book is that humour is essential to social life, and it's an obvious thing to say that humour is not a singular entity. For anyone interested in humour, the nature of it makes it harder to analyse than other social phenomena, as it is so idiosyncratic. The use of humour encapsulates many different aspects of what it is to be human. People do not react to humour in the same way and humour can be used as an excuse for unacceptable behaviour or a positive way of managing stress and establishing relationships.

This is the first book which has focussed exclusively on humour use in relation to social work, in workplaces, by service users and in fiction, so this book makes a unique contribution to social work thought.

I found that it is the underlying and often unconscious social mechanisms which allow humour to flourish and it is possible that humour could be part of the skills in a social worker's toolbox for working with people. Rather than focus solely on the defects people experience in their lives, humour can be used to think differently and innovatively and in the end positive and shared humour enables a practice that can be both more social and more humane.

Is there such a thing as social work humour?

There is pressure perhaps to engage with the subject of humour and 'we belong to a society in which fun has become an imperative and humour is seen as a necessary quality for being fully human' (Billig, 2005, p. 13). Much of this book has been drawn from sources which are not directly related to social work literature and practices. This reflects the episodic and inconsistent interest the profession has shown in relation to humour. Social workers are often alongside very unhappy people, who are experiencing trauma and immense personal difficulties. These people do not need their unhappiness to become the object of humour, although at times humour can be unacceptable and oppressive, used by some to minimise service users' distress or to create distance between the service user and the social worker, it is the contention of this book that even the uncomfortable problematic aspects of humour are also worthy of examination, because humour exists and is unavoidable.

So is there such a thing as social work humour? In short, I believe there is and this can be found in the work of Fran Orford, Harry Venning and in the TV series *Damned* with their attempts to make practitioners entertaining and interesting and the practice of the profession more visible to a wider public. I believe ironically that this demonstrates a profession that is growing in popular consciousness, despite the evidence that the future looks concerning for social work and public services as a whole.

The future of social work

Given that social work is conditioned by the societal context from which it emerges, and given the centrality of humour in contemporary society, it is possible to hypothesise that an exploration of humour, and the jokes which are told about social work, provide some insight into the role and place of social work in contemporary British society. It is right to have some anxieties about the outlook for the social work profession and 'we remain, for the most part, a diffident, under-confident and uncertain bunch in our dealings with the world beyond our own professional community' (Cooper, 2018, p. 8).

The social work profession is probably more anxious for its future survival now than at any other time as it is faced by a government that is both hostile to social workers and to children's services:

> *its message is an assumption that the present generation of social workers, and the people who teach them in universities, are failing. Just as in education, the Tory answer includes a pilot scheme in a number of English authorities to create Academies of children's services – in effect the beginning of privatisation models for an area of the welfare state that was taken over from the voluntary sector generations ago in order to achieve consistency and quality of service. if successful, it will have virtually taken children's social work services from local authority control and placed it in the hands of other organisations like Richard Branson's Virgin Care, who are concerned solely with profit.*
>
> (Turbett, 2018)

A profit driven privatised service for protecting vulnerable people is not something which sits comfortably with the profession's ethics and values.

Max Siporin who was a famous American academic, practitioner and writer was one of the first social workers to write about humour and noticed that social workers are more often likely to avoid joking about their work in public, perhaps because they lacked the degree of self-confidence and self-esteem to make light of themselves and their work and it's likely that the depressed state of many of their clients often rubs off on the workers (Siporin, 1984). These are serious and troubling times, so in this context what is the value of examining humour about social work? I want a future for social work in which it is as valued by everyone as much as I value it. I remain wedded to the idea that social work, the raw ideal of 'doing good' remains a wholly admirable ideal to commit to and that social work has achieved much. In this sense I think social work, much like humour, cannot be smothered, and the profession can take heart from Ritchie's words that social work like 'the comic spirit may have been suppressed but it can never be extinguished' (Ritchie, 2010, p. 159). I think there will always be a need for people who will try to do good and help others and these people will be proud to be called social workers.

The creativity of humour and contradictory social realities

Social work and social workers often find themselves in a contradictory position in society. Social workers are required by their standards of practice and proficiency, their code of ethics to treat the people they care for with kindness, care and respect, yet those people can be cruel, abusive and violent. Social workers practice in a world which places contradictory demands on them and 'is a profoundly ambivalently positioned and valued activity socially. Why? Well surely because we constantly threaten to "bring to the surface" – bring to mind – news of deeply disturbing and painful truths and realities about the lives and predicaments of people in our social midst who are suffering, dying needlessly, traumatized, traumatizing, displaced, abused, abusing, neglected, neglectful' (Cooper, 2018, p. 221). Humour has an important role in relation to social workers as it enables social workers to hold contradictory views (about their work, their service users and their organisations) at the same time. Given the contradictory subversive character of social work itself the two appear to me to go hand in hand, i.e. society manages its ambivalent and contradictory view of social work through jokes made about social workers, and social workers manage their ambivalent/ contradictory views of social work and service users though their use of humour. Winnicott (1953) in his work on transitional spaces argued that humour provided a space between the child's imagination and the real world outside. In a similar way social workers' use of humour and jokes permits them to talk about real and unreal situations. As a result potentially grave situations become not only less threatening, but amusing. Humour allows social workers to simultaneously live and practice in two worlds. Extending Winnicott's idea one could argue that it is almost as if humour allows social workers to negotiate and occupy the space between the true and the false self.

It seems to me that for social work to be successful in its endeavours it must do contradictory things and has attempted to do these things, so for example it has to care for people who often do not want the care or intervention, or it must challenge the state even though it is an arm of the state. Koestler (1964) saw humour as a creative device useful in understanding complex and contradictory social realities and given that social work is about dealing with complex and contradictory social realities, it is appropriate to put the two together.

Dullness v humour

In 2012 one local authority had as its tag line that they and their practitioners are 'serious about social work.' Whilst this appears to be a worthy gesture, this implies that any examination of humour in relation to social work runs the risk of not taking the practice of social work seriously, after all could you imagine the tag line 'humourous about social work,' or would other professions feel the need to emphasise their earnestness and say for example 'serious about nursing'? It could be argued that this tag line itself reflects a desire to be taken seriously from a profession which lacks confidence, which fears ridicule or the implication that social work is not an earnest endeavour. This I suggest must be understood within the context of social work's own insecurities, anxieties and fear about whether it is taken seriously, not just by its service users, but by wider society. This is an important issue, as without its own and society's 'legitimising' support, how can the social work profession be trusted to make the very difficult interventions it is required to make?

Clement Attlee in his book entitled *The Social Worker* which was originally published in 1920 articulated what he thought was a profound misconception about social workers: 'social workers, someone will say rather pityingly,good people no doubt in their way, but very dull' (original 1920, reprint 2018 Attlee, p. 2). I find myself agreeing with Attlee that this is a misconception and that social work is anything but dull. Attlee also characterised social work as an 'expression of the desire for social justice, for freedom and beauty and for the better apportionment of all the things that make up the good life' (Attlee, 2018, p. 2). When I conducted my research with social workers I began by asking them what was the funniest thing that had happened to them. One interviewee responded by talking about his experience:

> *Funny afterwards I suppose, rather than at the time but I was chased down the road with an axe . . . definitely not funny at the time; but being chased out of the flats with a consultant psychiatrist running faster than I could was quite amusing but . . . the consultant tried to get out quicker than I could, . . . he was scared of dogs as well and the guy had an American pit bull. . . . He had a pit-bull in one hand and a pickaxe in the other . . . at the time it was scary, but afterwards when you sit back you can see the funny side of it.*

In part this book has been an attempt to challenge the perception of social work as a 'very dull' profession and it's not possible, I would posit, that a good life, as Attlee put it, could not happen without the presence of good humour. You have to admire a practitioner who can reflect on a dangerous and frightening experience and sanitise that experience through humour and continue to practice. People might accuse the practitioner of bravado, but the comment shows that social work is anything but dull, and as Charlie Chaplin once said, 'to truly laugh, you must be able to take your pain and play with it' (Goodreads, 2018).

The 'social' aspects of social work

Social workers like to use humour, as do people who experience services and this itself is a unifying social reality as humour is a universal phenomenon experienced by human societies across time and across cultures, although what people find humourous changes with time and context. We all still laugh and find humour in the bleakest of times. It would appear therefore that social workers who use humour and jokes are trying to do several 'risky' things at the same time, to show their human characteristics, discharge their own emotions and establish rapport. All of which it could be argued is ultimately about establishing relationships with service users, their colleagues and their managers. Added to this humour also helps social workers develop their resilience, as humour is closely tied to resilience *particularly the capacity to laugh at oneself* (Furnivall, 2011).

Social workers sometimes have feelings of fear or anger towards their service users, but must treat these service users with respect and integrity. As a result social workers need to find ways of managing the competing and complex feelings they have towards their service users. Humourous or funny things happen all the time in social work, and social workers often talk about them, even in the midst of more overwhelming sadness and distress. It is insensitive and inappropriate to ridicule and make fun of people who are in vulnerable and unhappy situations yet ridiculing them may help them to gain some insight into the stupidity of their actions or decisions. However, social workers also know that in talking in such a way this can go against the core value base of social work. So how do workers manage this contradiction? They displace their emotions through the use of humour with each other. In my experience service users often made social workers laugh, and whilst some of this humour was shared at other times humour was made at the service user. No social worker would argue that this is ethically acceptable, but social workers have a need to safely discharge their negative feelings about service users, sometimes in the absence of reflective supervision or the absence of management surveillance.

One aspect of this book has been to recapture what it is that makes social work 'social' in the sense that it about relating to others and making social connections with other people, one social worker said to me once that there was nothing which showed your humanity more than humour. It is possible to suggest that being viewed as humourless is far more dangerous for the future of the social work profession as it makes social work more vulnerable to attacks from those who wish to undermine its role in society.

Frost (1992) found that service users needed to be given the opportunity to share their stories as these often revealed great humour in how they managed their lives and suggested that humour could be used in working with service users with such questions as: *tell me about the last time you had a good laugh* or *would you tell me your favourite joke* could be posed as part of an assessment.

In conclusion there is no single definition of the relationship between social work and humour, however there are key features, and this book shares Lockyer and Pickering's assertion that humour 'infiltrates every area of social life' (Lockyer and Pickering, 2009, p. 6.) The implication of this infiltration is that humour insinuates itself into social work practice and as such is unavoidable. For this reason social workers have to find ways of personally and professionally managing humour, as it is unescapable. As a society we fear the word joke as to be seen as a joke is undermining to any person or to a profession. Social workers fear not being taken seriously, but conversely use humour and jokes to manage their unhappiness at work, or to cope with the stress of the work. In the end humour is ever present and understanding it and engaging with it is crucial for practice as Bollas said 'a sense of humour – which takes pleasure in the contradictory movement of two objects . . . [means] we may choose between our comic and tragic potential' (Bollas, 1995, p. 245).

References

Attlee, C. R. (2018) *The Social Worker* London: Forgotten Books (original London: G. Bell and Sons Ltd. 1920).

Billig, M. (2005) *Laughter and Ridicule towards a Social Critique of Humour* London: Sage Publications.

Bollas, C. (1995) *The Work of Unconscious Experience* London: Routledge.

Cooper, A. (2018) *Conjunctions: Social Work, Psychoanalysis and Society* The Tavistock Clinic Series London: Karnac.

Frost, C. (1992) *Having Fun in Social Work* Middle Tennessee State University Paper http://capone.mtsu.edu/cfrost/soc/thera/HUMOR.htm (accessed 24/8/13).

Furnivall, J. (2011) Guide to developing and maintaining resilience in residential child care *Community Care Inform Article*.

Goodreads (2018) *Charlie Chaplin>Quotes>Quotable Quote* www.goodreads.com/quotes/245052-to-truly-laugh-you-must-be-able-to-take-your (accessed 1/8/18).

Koestler, A. (1964) *The Act of Creation* New York: Palgrave Macmillan.

Lockyer, S. and Pickering, M. (Eds.) (2009) *Beyond a Joke: The Limits of Humour* Basingstoke: Palgrave Macmillan.

Ritchie, C. (2010) Against comedy *Comedy Studies* 1(2), 159–168.

Siporin, M. (1984) Have you heard the one about social work humor? *Social Casework* 65, 459–464.

Turbett, C. (2018) Why social work is under siege thanks to the UK Government *The CommonSpace* www.commonspace.scot/articles/3318/colin-turbett-why-social-work-under-siege-thanks-uk-government (accessed 30/7/18).

Winnicott, D. W. (1953) Transitional objects and transitional phenomena: A study of the first not-me possession *International Journal of Psycho-Analysis* 34, 89–97.

Index